Construction Sealants and Adhesives

Wiley Series of Practical Construction Guides

M. D. Morris, P.E., Series Editor

William G. Rapp
CONSTRUCTION OF STRUCTURAL STEEL BUILDING FRAMES

Jacob Feld
CONSTRUCTION FAILURE

John Philip Cook
CONSTRUCTION SEALANTS AND ADHESIVES

Construction
Sealants and Adhesives

JOHN PHILIP COOK

Jacob Lichter Professor of Engineering Construction
University of Cincinnati

WILEY-INTERSCIENCE, A Division of John Wiley & Sons, Inc.

New York · London · Sydney · Toronto

Library of Congress Catalogue Card Number: 70-121905

ISBN 0 471 16900 5

Printed in the United States of America

10 9 8 7 6 5 4 3 2

Series Preface

The construction industry in the United States and other advanced nations continues to grow at a phenomenal rate. In the United States alone construction in the near future will exceed ninety billion dollars a year. With the population explosion and continued demand for new building of all kinds, the need will be for more professional practitioners.

In the past, before science and technology seriously affected the concepts, approaches, methods, and financing of structures, most practitioners developed their know-how by direct experience in the field. Now that the construction industry has become more complex there is a clear need for a more professional approach to new tools for learning and practice.

This series is intended to provide the construction practitioner with up-to-date guides which cover theory, design, and practice to help him approach his problems with more confidence. These books should be useful to all people working in construction: engineers, architects, specification experts, materials and equipment manufacturers, project superintendents, and all who contribute to the construction or engineering firm's success.

Although these books will offer a fuller explanation of the practical problems which face the construction industry, they will also serve the professional educator and student.

M.D. MORRIS, P.E.

Preface

This book deals with the materials and methods of sealants and adhesives in construction. It is intended for architects, engineers, contractors, and young men studying in any of these fields. It is not my intention to treat the complicated chemistry of sealants and adhesives. Practicing engineers and architects are interested in the performance of a material, not its chemistry. This book therefore attempts to assemble the available information on the performance of these materials and set it down in an organized form so that it can be used.

The book is organized so that the reader is led from source, to need, to design, to properties, to installation, and then to specific materials and their properties.

The portion of the text on adhesives is understandably brief. Although many adhesive products are used in construction, there are relatively few instances in which the architect or engineer actually specifies the adhesive to be used.

It would be impossible to thank everyone who has helped in the production of this book. Virtually every major manufacturer has contributed in the form of pictures, samples of material, and previously published papers. Special thanks are due to the Darworth Division of Ensign-Bickford Corporation, Products Research and Chemical Company, and the Tremco Manufacturing Company for their help with pictures and written background material.

I also acknowledge the help of Mr. Harold B. Britton of the New York State Department of Transportation and also the Thiokol Chemical Company for their help over the years. Thanks are also due to Mr. Thomas Via for the hours he spent preparing many of the drawings for this text.

John P. Cook

Cincinnati, Ohio
March 1970

Contents

Construction Sealants and Adhesives

1

Introduction to Sealants

The last 20 years have seen a rather complete change in the nonresidential construction methods used throughout the world. The relatively low-rise bearing wall-type construction has given way to frame construction, which erects a building skeleton of steel or concrete. This framework is then wrapped in a separate envelope or curtain wall. The wall envelope may be of concrete panels, unit masonry, glass or metal panels, or other materials.

These structures are inherently more flexible than bearing wall construction and present new problems in weatherproofing the structure. Paralleling the growth of curtain wall construction has been the growth of a whole new family of higher-performance specialty caulking, or sealing compounds. The term "sealant" was first used to differentiate these new sealing compounds from the older oil-based caulks. Modern usage, however, has extended the meaning of the word, so that the term "sealant" is now used to include all the types of weatherproofing joint materials currently in use.

The meaning of the term "sealant" has been broadened by usage so that it now includes viscous liquids, mastics or pastes, tapes, and gaskets. Sealants may thus be considered as any material placed in a joint opening, generally for the purpose of weatherproofing the building. They are designed to prevent the passage of moisture, air, dust, and heat through all the joints and seams in the structure.

Because of the wide variety of requirements imposed on the various sealing materials, they naturally take on different forms.

Viscous liquid sealants are prepared in a pourable form, generally for use in horizontal joints, as in patio and pool decks, sidewalks, or roof decks. Because of the flow characteristics, they generally tend to fill any voids in the substrate and to provide better wetting action and, consequently, better adhesion than the gun-applied sealants.

The mastics, or gun-applied sealants, contain a thixotropic agent to control flow and prevent the material from sagging or flowing out of the joint. These materials are generally installed with a standard caulking gun, but they may be applied by knife or trowel.

The tapes may be either cured or uncured, and are used as bedding compounds and also for architectural glazing work. The tapes have the obvious advantage of being furnished in roll form and thus are very easy to apply. They can be butt spliced on the job, and their roll form enables glazing of a structure from the inside, which saves considerable time and labor.

The cured gaskets may be either extruded or foamed, and are available in a multitude of shapes and forms for various sealing jobs. They are installed by forcing the gasket into the joint opening, and depend on the pressure of the gasket against the joint to maintain a tight seal. Consequently, they are generally fabricated from high recovery elastomers. They are somewhat more expensive than the tapes, but they are often preferred because of the wide variety of available shapes and the excellent finished appearance of the joint.

1.1   The Market

For the past several years, total new construction activity has maintained a rather steady 10% of gross national product (GNP). Of this total construction volume, roughly 30% is devoted to nonbuilding construction (bridges, highways, and dams) and 70% is devoted to building construction. Of this building construction volume, 55 to 60% is residential construction, whereas 40 to 45% is nonresidential construction.

During the last decade, the new construction industry has grown approximately 4% per year. The total sales of sealant products, however, have grown approximately twice this rate and are expected to maintain this trend. In addition to the new construction market, the total construction costs for maintenance, repair, and remodeling in 1966 amounted to approximately $30 billion. Of this amount, $13 billion represents nonresidential work and $17 billion represents residential work.

Together, the new construction and maintenance work in 1966 represented for the sealants industry a total sealant and caulk consumption of 2.4 million gal valued at $24.3 million for the commercial market, and 92.4 million cartridges valued at $21.2 million for the largely residential,

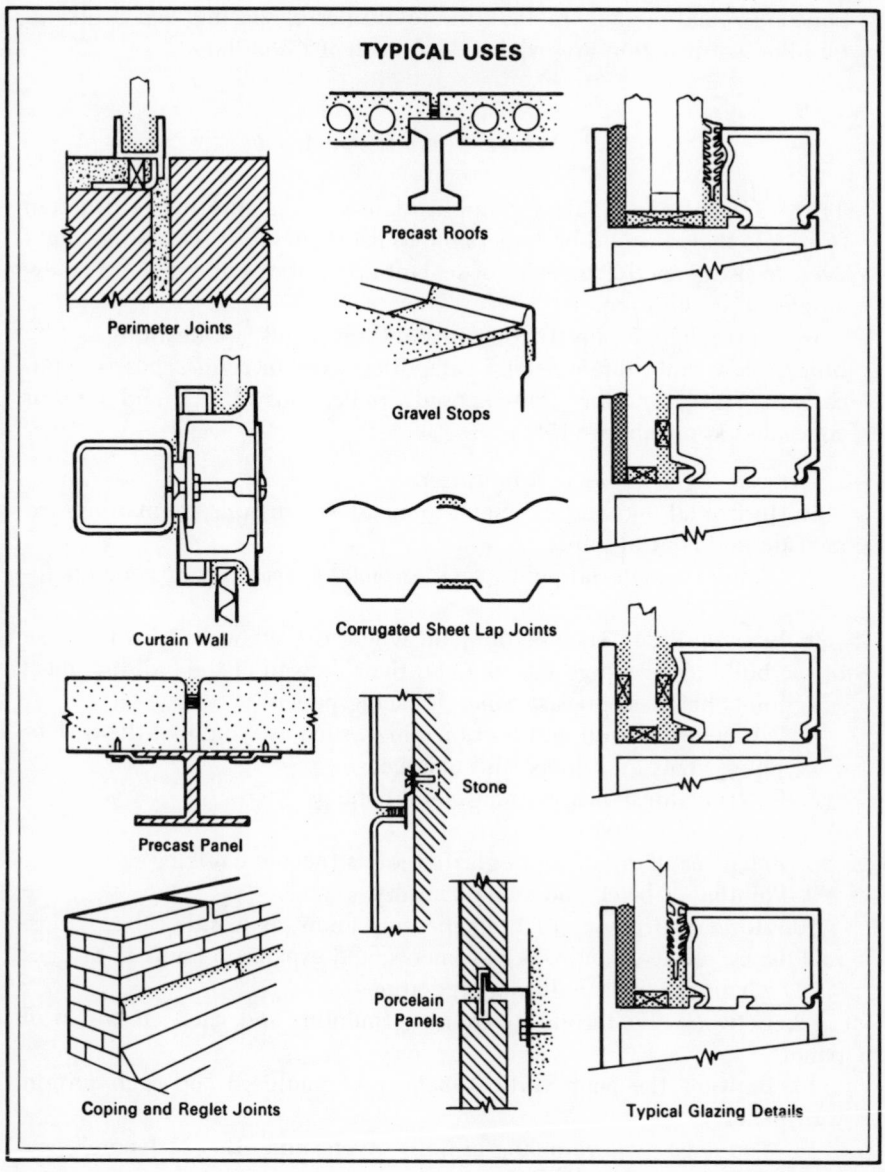

TYPICAL USES

Perimeter Joints

Precast Roofs

Gravel Stops

Corrugated Sheet Lap Joints

Curtain Wall

Precast Panel

Stone

Porcelain Panels

Coping and Reglet Joints

Typical Glazing Details

Fig. 1.1 Typical sealant applications (Courtesy of Tremco Manufacturing Company)

noncommercial market. In 1972 the total sales of sealants and caulks in building construction are expected to reach $62.3 million.

1.2 Where Sealants Are Used

There are so many different applications for sealants in construction (Fig. 1.1) that it would be impossible to list them all. It is possible, however, to list general categories of sealant applications and give a few examples in each group.

In nonresidential construction, sealant and caulk applications can be broken down into three general categories: exterior sight-exposed applications (70%), interior sight-exposed applications (15%), and interior concealed applications (15%).

Exterior Sight-Exposed Applications
1. Horizontal and vertical metal-to-metal and masonry-to-masonry expansion and control joints
2. Dissimilar material joints, such as metal-to-masonry or concrete-to-wood
3. Junction joints, such as the joint where the sidewalk joins the base of the building or where the roof and the side wall of the building meet
4. Joints between precast concrete facing panels
5. Perimeters of wall and roof openings, such as windows, doors, louvers, pipes, vents and ducts, and chimneys
6. Exterior thresholds, flashings, and copings
7. Pools and reservoirs
8. Repair caulking of small glazing joints (needle caulking)
9. Pointing of brick and stone masonry
10. Horizontal paving or traffic-bearing joints, as in sidewalks, patios, roof decks, terraces, highway pavements, and expansion joints in bridges
11. Channel and face glazing operations
12. Between the secondary framing (muntin) and each curtain wall panel
13. Between the main vertical metal post (mullion) and each curtain wall panel
14. Tappings (waterproofing of bolts, rivets, and other fasteners)
15. Glass panels for skylights
16. Corrugated metal roofing and walls
17. Spandrels

Interior Sight-Exposed Applications
1. Perimeters of doors and fixed window frames
2. Exposed interior masonry control joints

3. Joints and recesses between metal frames and interior masonry
4. Interior construction joints in concrete and plaster
5. Joints at interior masonry or plaster walls which adjoin columns, pilasters, or exterior walls
6. Wood-to-masonry joints
7. Joints where a precast concrete ceiling meets a wood partition
8. Interior duct work
9. Plaster and metal trim joints in curtain wall buildings
10. Expansion or control joints in industrial floors

Interior Concealed Applications
1. Joints in exterior walls
2. Concealed masonry-to-floor structure joints
3. Exterior and interior metal thresholds, saddles, and sills
4. Joints and recesses between access panels, electric panels, piping, and pipe sleeves
5. Between metal frames and masonry substrates as in show windows
6. Between overlaps of mating or interlocking metal joints in curtain wall construction
7. Between mullions and curtain wall panels

In the residential market, the builder and painting contractor use sealants for perimeters of wall and roof openings such as doors, windows, chimneys and vent stacks, and for glazing work. The homeowner uses caulks and sealants for these same applications as well as for filling cracks and sealing tubs and shower stalls. Approximately 80% of residential uses are exterior.

1.3   Who Supplies the Sealants

This industry is characterized by raw material suppliers who furnish basic ingredients to a large number of formulators. These raw material suppliers include many of the major chemical companies in the United States. They supply basic polymers or oils to approximately 100 formulators of medium to large size, who cover a fairly broad geographical range, and an equal number of compounders who operate strictly on a regional basis. Accessory ingredients, such as fillers, plasticizers, color pigments, and adhesion additives, are furnished by a wide range of suppliers.

The result of this dispersal is that the quality control of sealant products depends on a multitude of formulators whose expertise would naturally vary quite widely.

There are two notable exceptions to this general system. The manufacturers of silicone sealants (General Electric Company and Dow Corning Corporation) market their sealants under their own trademarks. Thiokol Chemical Corporation, which furnishes the basic polymer to virtually all formulators of polysulfide sealants, has initiated a Seal of Security program under which they test the finished products of their formulators to insure certain performance levels.

1.4   Who Specifies the Sealants

In the noncommercial or consumer market, the contractor is the primary buying influence on new construction. Approximately 75% of the time he specifies not only the type of product, but also the brand.

In the repair and maintenance market, the small painting contractor and the homeowner are the major buyers of sealant products.

In heavy nonbuilding construction, the engineer specifies what sealants will be used. This type of construction includes highways, bridges, dams, airports, and water supply and sewer construction.

In new nonresidential construction, the sealant and caulking compounds are essentially the choice of the architect. Nonresidential construction includes educational, industrial, commercial, religious, institutional, social, recreational, and manufacturing buildings. In addition, the architect is the primary influence in *major* nonresidential remodeling and in *custom* residential homes (which represent only about 15% of housing starts).

Manufacturers of certain prefabricated units, such as windows and panels use a sizable segment of sealant materials, but these companies exercise their own quality control and normally furnish a warranty on the finished unit.

1.5   How Sealants Are Specified

There is no set pattern in the building industry by which sealants are specified. The approach to specifying a sealant necessarily varies according to the size and type of the project. Listed below are some of the approaches used to specify a sealant material.

1. Specify reference standard (e.g., meeting certain Federal specifications)
2. Specify a type of sealant (e.g., polysulfides) and also name two or three suppliers
3. Prescribe the performance (physical/chemical properties) required

of the sealant and name two or three commercial products which meet this performance

4. Specify the products of one particular company
5. Name the specific trademark products of several suppliers
6. Name a specific product (e.g., two-component polysulfides)

In nonbuilding construction, especially bridge and highway work, much of the work is funded by the individual states. Most states have complete construction standards and specifications based on Federal or ASTM ° specifications or test methods.

In public building construction, the architect is somewhat limited in his specification writing by mandatory open bidding. In this type of work, the architect generally must include an "or equal" clause in the specification. This means that the architect may not limit acceptance to certain proprietary materials. Although the bidder may substitute any material which meets the performance requirements of the specification, he must obtain the architect's approval of his "or equal" material. In private building construction, however, the contractor is overruled by the architect and must follow his specifications closely.

The architect is therefore the person held responsible for the sealants and caulks he specifies. Because of this liability, architects naturally tend to be conservative in their choices and use those materials with which they have had a successful record of performance. On the other hand, for nonspecified private remodeling work, the caulking contractors are generally concerned with nonworking joints and are justified in using the lowest priced materials available.

The architect will frequently hold the sealant manufacturer responsible for the performance of his material, and will require that a manufacturer's representative be present at the job site before and during installation. Moreover, many architectural firms will require that the manufacturer offer a performance guarantee for a specified time period and will even insist that the manufacturer pay for a complete recaulking job if the building leaks.

In order to consider a sealant for use on a building, architects often insist on complete technical data sheets from the manufacturer. These "tech data" sheets contain information about the following properties of a sealant:

1. Pot life
2. Cure time
3. Mixing instructions (if required)
4. Modulus of elasticity

° American Society for Testing and Materials.

5. Ultimate elongation
6. Temperature limitations

The data sheets should also indicate which Federal Government or USASI ° specification the material meets. Naturally, these specifications are considered as minimums by the industry; hence the data sheets usually indicate to what extent the material exceeds these specifications. Most architects also insist that the supplier furnish results of roughly three years of field experience before specifying a new material for a major project.

Representative costs for specification-type sealants are given in Table 1.1.

1.6   Specifications

In putting together the contract drawings for a structure, the architect writes up a set of specifications covering materials and their installation. It is common practice for the architect to refer to other standard specifications and to include them as a part of his document. For example, his specification might state, "The sealing compound for the joints in the precast wall panels shall be a one-component sealant conforming to the requirements of Federal Specification TT-S-00230." In essence, then, the architect uses certain standard specifications as guides and may, at his discretion, add other requirements or otherwise modify the standard specifications to suit the performance requirements of a particular structure.

The writing of the standard specifications was itself a somewhat reverse process. It would seem logical to define the properties required of a material strictly on a performance basis and leave it up to the sealant manufacturers to meet these requirements. Unfortunately, this was not the case. In most instances, a material giving fairly good performance was used as a model, and a specification was written describing the properties of this material. Thus, although the language of the specifications was made as broad as possible, the standard specifications do tend to favor one material.

At the present time there are at least 10 groups working on new sealant specifications or upgrading older ones. Considering all the sources, there are at least 100 specifications for various types of sealing materials. The Compendium of Sealant Specification Abstracts, published by the U.S. Government, lists over 30 specifications. Out of this maze of pub-

° United States of America Standards Institute.

lished information, very few specifications have received enough acceptance to be considered as standards by the sealants industry.

For the viscous or mastic sealants, there are four standard specifications in widespread use for building construction work:

Federal Specification TT-S-00230 is used for a sealing compound, synthetic rubber-based, one-component, and chemically curing. The single component polysulfide sealants meet this specification, as do the single component urethanes, silicones, and some of the solvent-based acrylics.

Federal Specification TT-S-00227 is used for a sealing compound, rubber-based and two-component, and is furnished in two grades. The flow (self-leveling) grade material is for use in horizontal joints such as sidewalks and patios, and the nonsag grade is for use in vertical or overhead surfaces. Both polysulfides and urethanes meet the requirements of this specification.

Federal Specification TT-C-598 is intended for oil- and/or resin-based caulks. This specification has largely superseded the older TT-P-791a which was originally intended for wood sash glazing. Most of the higher quality oil-based caulks will meet the requirements of TT-C-598.

USASI 116.1 is an upgrading of the older ASA ° 116.1, which was written for polysulfide-based sealing compounds. This is not a Federal specification; until its upgrading it had been largely replaced in actual usage by TT-S-227.

In 1968 the National Association of Architectural Metal Manufacturers published a specification for noncuring compounds, such as the butyls, together with specifications for sealing tapes. However, these specifications have not received the widespread acceptance they probably deserved and are not generally considered as standard specifications.

Thus the standard specifications cover the elastomeric sealants and the oil-based caulks, but do not cover such other materials as butyls, latex-type caulks, or preformed tapes or gaskets. Various ASTM test methods cover certain properties of these materials, but the ASTM methods would not, in general, be broad enough to find widespread usage in architectural specifications. ASTM is currently working on a specification for the preformed gaskets.

All of the standard specifications have drawbacks, and several industry groups are actively working to upgrade the specifications and make them more effective. Especially active in this area have been ASTM and the Society of the Plastics Industry.

° American Standards Association.

Table 1.1 Typical Sealant Applications and Representative Cost

Sealant Base	Cost Range (dollars per gal)	Uses
Nonhardening sealants		
Oleo-Resin	2–6	Sealing of concrete joints, masonry copings, glazing and sealing of cable pressure splices, electrical conduit, and glass-to-metal meter cases.
Asphalt and bituminous	1–4	Sealing faying surface metal joints, silos, and air-conditioners. Caulking for expansion and contraction joints.
Butyl	5–12	Caulking expansion and contraction joints. Metal-to-glass seals, and metal-to-metal sealing to separate dissimilar materials. Sealing electrical conduit.
Acrylic	8–12	Pipe joints, glazing, masonry, and metal caulking. Special compounds used as liquid gaskets and pipe dope.
Polybutene	3–7	General construction-type caulking and glazing. Seal between dissimilar metals.
Hardening sealants—rigid types		
Epoxy	7–12	Potting electrical connectors, encapsulating miniature components, coating circuit boards, and cable splicing. Caulking and pipe sealing. Used as a sealer and abrasion-resistant coating for concrete.
Polyester	6–12	Potting, molding, and encapsulating. Gasket and pipe thread sealants.
Oleo-resin	2–6	Same general uses as nonhardening formulations. Hardening type materials
Asphalt and bituminous	1–4	should be used where pressure limits exceed the limitations of the nonhardening formulations.

Hardening sealants—nonrigid types

Chemical-reaction systems

Type		Description
Polysulfide—two-part	10–22	Sealing integral fuel tanks and pressure tanks. Sealing faying surfaces and channels. Potting, molding, and sealing of plexiglass. General construction sealing and caulking of metal, wood, and masonry joints.
Polysulfide—one-part	20–25	General construction-type sealing, caulking, and glazing. Sealing between dissimilar metals. Deck caulking and sealing of refracting mirrors.
Urethene—two-part	12–15	Potting and molding of electrical connectors, encapsulation of hydrophones, transducers, and circuit boards. Caulking where compatibility with LOX is required.
Urethane—one-part	12–18	General construction-type sealing, caulking, and glazing.
Silicone—two-part	45–100	Potting and molding of electrical connectors. Potting firewall connectors, and coating umbilical cables. Sealing of heat shields.
Silicone—one-part	24–37	General construction-type sealing, caulking, and glazing. Potting and molding.
Polysulfide—modified Bitumens—two-part	4–6	Sealing of expansion and contraction joints in runways and taxi strips where flame and fuel resistance are required.
Modified epoxy—two-part	7–12	Potting, molding, encapsulating. Sealing transformers. High-voltage splicing, capacitor sealing. General construction caulking.
Acrylic—one-part	8–12	General construction-type sealing, caulking, and glazing.
Viton—two-part	80–115	For high-temperature service where fuel and oil resistance is required. Sealing fuel tanks, channels, and faying surfaces.

Solvent-release systems

Type		Description
Neoprene	8–12	Sealing between dissimiliar metals. Caulking and general sealing.
Hypalon	8–12	Similar in uses to neoprene.
Butyls	5–12	Glazing and caulking of metal, glass, and masonry-type joints.

Reprinted by courtesy of *Machine Design*.

Interestingly enough, for private construction work, there is actually no such thing as "obtaining approval" for a sealant under a Federal specification. The specification contains certain requirements which the sealant must meet. Hence the manufacturer must obtain copies of the specification and either conduct the tests in his own laboratory or have them performed by an independent laboratory. As soon as the manufacturer is satisfied that a material meets the requirements of a specification, he may claim so in his promotional literature.

Most architects are aware of the limitations of the standard specifications and, consequently, they tend to accept the manufacturers' technical data sheets with a certain degree of caution. The majority of architects will insist on a proven record of field performance before accepting a material for use on a given project. However, a sizable minority of architects feel that accelerated laboratory testing by a reputable, independent laboratory will suffice. Some well-known and established laboratories with a capability in this area are the following:

> Vjorksten Research Laboratories, Inc.
> Madison, Wisconsin
>
> Battelle Memorial Institute
> Columbus, Ohio
>
> Chemical & Engineering Associates, Inc.
> Elkton, Maryland
>
> Colburn Laboratories, Inc.
> Chicago, Illinois
>
> Pitt Testing Laboratories
> Pittsburgh, Pennsylvania
>
> Armour Research Foundation
> Chicago, Illinois
>
> Foster D. Snell, Inc.
> Cambridge, Massachusetts
>
> United States Testing Co.
> Hoboken, New Jersey

If a manufacturer wishes to sell his products directly to the Federal Government, the entire specification picture changes. The manufacturer must complete a Bidder's Mailing List Application and submit his bid on the particular job. The purchase by the Government is made from the low bidder. Before the actual award is made, however, a Government inspector is sent to the plant to insure that the manufacturer has the necessary production facilities. Moreover, for certain products there is a "quality products list" (QPL), whereby only the bids of certain approved manufacturers will be accepted upon submission. In order to be placed on this list, a company must submit samples of its products for analysis and approval by the Federal Government. The U.S. Government maintains several excellent laboratories which are well equipped and staffed with competent personnel. Building sealants are evaluated at the National Bureau of Standards Laboratory in Washington, D.C. Highway and airport paving sealants are tested at the Ohio River Division Laboratory of the Corps of Engineers in Cincinnati. Asphaltic materials are evaluated at the Waterways Experiment Station of the Corps of Engineers in Vicksburg, Mississippi. Sealants for canals, dams, and other reclamation projects are handled by the Bureau of Reclamation in Denver, Colorado.

2

Weatherproofing the Building

2.1 The Need for Sealant Materials

The exterior of a building must be weatherproofed in order to prevent drafts and wind-driven rain from entering the building. With today's emphasis on year-round air-conditioning, it is of paramount importance that the building be properly sealed in order to maintain this level of conditioned comfort. Water leakage into a structure may cause structural deterioration and usually presents unsightly damage and costly repairs. Far less dramatic, but just as costly in the long run, are the increased costs of heating and air-conditioning which can result from a poorly sealed structure.

Most buildings are well designed and properly sealed when they are new. However, movement of the building or its component parts often causes sealant failure and resulting leaks. This movement of the building might be caused by expansion and contraction of its components, by wind load on the structure, or by settlement. Buildings of heavy bearing wall construction tend to remain fixed and absorb any movement over a large number of joints. In the heavy brick wall, for instance, building movement may show up as fine hairline cracks in the mortar joints which are not, in general, too susceptible to rain damage. On the other hand, the large, flexible structure with relatively few pieces, characteristic of today's curtain wall construction, may move a great deal. The joints between large precast panels may move ¼ in. or more due to expansion and contraction. During a high, gusty wind, the top of a large, flexible curtain wall building may move as much as 12 to 14 in. out of plumb. Naturally, such severe movements place a great deal of strain on any joint sealing material.

2.2   The Mechanism of Rain Penetration

In order for rain to penetrate the wall of a structure, there must be water impinging on the wall of the building, an opening in the wall through which the water can travel, and a force to drive the water through the opening.

During a driving rainstorm the water forms a film on the exterior surface of the building. This film of water flows down and covers the joints in the building surface. If there are any openings in the joints, the wind pressure acting on the surface forces the water into or through the building wall. Estimates vary as to how much water flows down a vertical wall. One research institute uses a figure of 40 gal / (100 sq ft) (hr).

Against an even wall surface the film of water maintains a rather constant thickness. However, projecting mullions tend to split the air flow and cause concentrations of water flow at the panel-to-mullion joint. Other projections on the building, such as cornices, also tend to cause turbulent air flow and water concentrations.

2.3   The Design of Exterior Wall Joints

There are two basic methods of weatherproofing the exterior wall of a building:

1. One-stage weatherproofing, which is the sealing of joints at the exterior face of the wall section (the wetted plane). Here the sealant material acts as both rain seal and air seal.

2. Two-stage weatherproofing, in which the air seal and rain seal are treated as separate functions. A deflector or rain shield is used to keep water out of the joint, and the air seal is placed in an interior, protected location.

Although each system of sealing has its advantages and disadvantages, any through-the-wall joint requires some form of seal. Since the functions of the seals are different in the two systems, the performance requirements of the seals also vary.

2.3.1   One-Stage Weatherproofing

Until very recently most American building practice has been based on the principle of one-stage sealing. Most repair and resealing work is also necessarily one-stage sealing. In addition, many sealant applica-

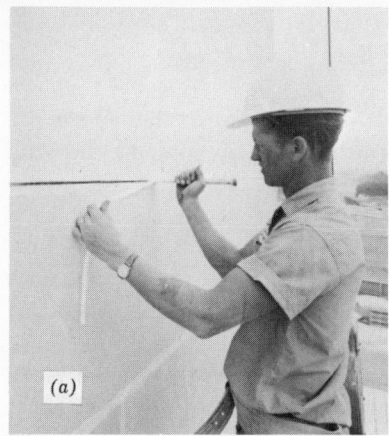 

Fig. 2.1 One-stage sealing. (*a*) Placing the back-up material. (*b*) Placing the sealant. (Courtesy of Tremco Manufacturing Company)

tions, such as the perimeters of wall and roof openings for louvers, vents and stacks, junction joints at the base of the building, lap joints in sheet materials, and horizontal traffic bearing joints, provide no other solution.

One-stage sealing is simply the elimination of all openings which would admit water past the exterior wall line of the building. In this method of weatherproofing the caulking contractor installs the sealant in the joint, using standard caulking methods. Thus, familiarity favors the one-stage seal. Figure 2.1 shows a typical one-stage seal and its installation.

The economics of sealing supports the one-stage seal simply because it is a one-step process. Another economic factor to be considered is the design of the edges of exterior building panels. The edges of panels, which form the interface of the joint, can be of a simple straight-line design (Fig. 3.1).

One-stage sealing, however, has several disadvantages which must also be considered. The sealant material is exposed to the rays of the sun which may cause damage to the sealant by ultraviolet radiation; the sealant and its substrate are subject to alternate wetting and drying; extension of the exposed sealant occurs when the building contracts in cold weather, and most sealants are less extensible when cold; also, the exposed sealant is subject to picking and gouging by vandals.

2.3.2   Two-Stage Weatherproofing

Two-stage weatherproofing has gained widespread acceptance in Europe and much of Canada, but has only very recently been considered in American building practice. This foreign growth has come not only because of the inherent merits of the system, but also because of economic factors. The higher priced sealants which have been used widely in this country are not as readily available in many foreign countries. Consequently, the Europeans and Canadians have developed joint designs which permit the use of lower performance (and price) materials.

Two-stage sealing treats the rain seal and the air seal as separate functions. The rain seal in horizontal joints can be achieved by a deflector or an offset in the edge of the building panel. (Fig. 2.2). In vertical joints, the rain shield is often a simple strip or tube of rubber placed in a recess in the edge of the panel (Fig. 2.3). The possibilities for forming this rain seal are endless once the designer recognizes that this seal should *not* be airtight. Its function is to prevent water from reaching the interior seal and thus from reaching the inside of the building. Two-stage sealing depends for its effectiveness not on completely preventing water penetration, but rather on controlling the forces that act to drive the water inward. Therefore, an important feature of two-stage sealing is the presence of a pressure-equalizing chamber behind the rain seal. (See Fig. 2.3). This pressure-equalizing chamber should be vented to the outside so that the air pressure in this chamber is the same as the external air pressure. Thus any wind-driven rain which does pass the rain baffle is trapped in this enlarged area and is drained harmlessly back to the exterior.

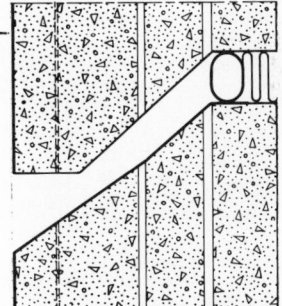

Fig. 2.2 Horizontal joint— offset panel edge with rain seal. (Joint design by Trygve Isaksen, Norwegian Building Research Institute)

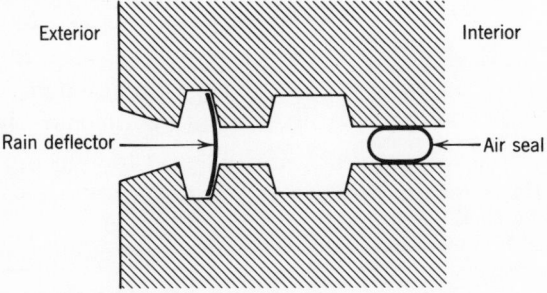

Fig. 2.3 Vertical joint with rain seal and air seal. (Joint design by Trygve Isaksen, Norwegian Building Research Institute)

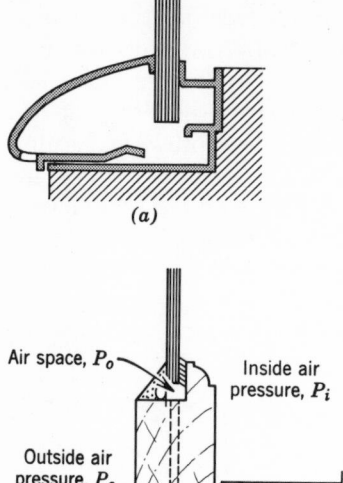

(a)

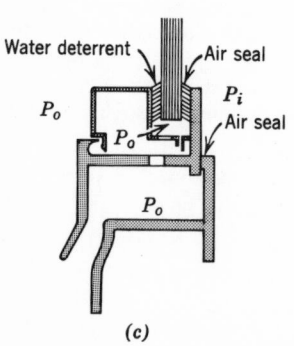

Air space, P_o Inside air pressure, P_i

Outside air pressure, P_o

P_o

(b)

Water deterrent   Air seal

P_o P_i

P_o Air seal

P_o

(c)

Fig. 2.4 (*a*) "Store front" glazing detail showing openings for air pressure equalization and drainage. (*b*) Single glass in wooden sash with openings for pressure equalization and drainage at sill. (*c*) Metal sash unit designed for pressure equalization. (Design by G. K. Garden, Canadian National Research Council)

Two-stage sealing is not new. In the broad sense of the term, American home builders have been using two-stage sealing for years. The general rule for insulating a residence is to insulate the inhabited areas and ventilate the rest of the structure. Thus the roof serves as the rain shield, attic areas are ventilated, and the air seal (insulation) is placed in a protected region between the ceiling joists.

Two-stage sealing therefore has the advantage of having the sealant material (the air seal) placed in a protected location. Thus the sealant is not exposed to ultraviolet radiation, extreme cold, or a wet-dry cycle. Under these circumstances the service life of the sealant may be substantially lengthened.

At first thought, it would seem impossible to apply two-stage sealing to glazing units. However, the Europeans and Canadians have developed successful designs (Fig. 2.4) which use the two-stage principle.

The two-stage seals also have some disadvantages. The sealant material is generally more difficult to apply properly in this enclosed position. Resealing in the event of failure may be extremely difficult. The designer must also consider the cost of forming the necessary recesses or offsets in the panel edges to accommodate the rain baffle and the pressure chamber.

2.4 *Types of Joints*

From a functional viewpoint, there are only two types of joints in building and construction: working joints which change size or shape with the relative movement of the adjoining parts and nonworking joints in which relative movement between adjacent parts is minimized or eliminated by the details of construction.

In low-rise buildings or buildings which cover large areas, the major working joints are isolation joints and control joints. The isolation joint, which is approximately 1 in. wide, divides the building into entirely separate sections. It is a through-the-building joint and is spaced at intervals of approximately 100 ft. The control joint, which is used primarily with masonry, is usually ⅜ to ½ in. wide. Formed as a chase or recess in the wall, it provides a weakened plane through the wall to confine or control cracking. The control joint is spaced at intervals of about 25 ft.

In high-rise curtain wall construction the major working joints are the joints between exterior facing panels or the panel-to-mullion joints. These joints become especially critical because they are usually sight-exposed and hence subject to weathering.

In working joints the sealant material has to withstand cyclic stresses over a long period of time without failure. In nonworking joints the sealant functions primarily as a filler and is subject to little or no stress.

Examples of Working Joints
1. Isolation joints
2. Control joints in masonry walls
3. Joints between exterior facing panels
4. Exterior panel-to-mullion joints
5. Junction joints, such as the joint where the sidewalk joins the base of the building
6. Horizontal joints in terraces, patios, and sidewalks
7. Lap joints in sheet roofing or siding

Examples of Nonworking Joints
1. Interior heel beads for glazing
2. Copings and gravel stops
3. Reglet joints
4. Sealing or glazing of small individual window lights
5. Preformed tape applications in glazing work

Most working joints can be classified as butt joints or lap joints. Butt joints subject the scalant to alternating tensile and compressive stresses,

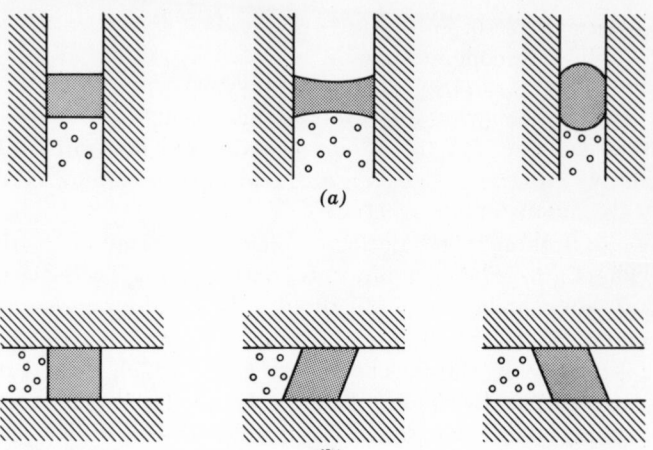

Fig. 2.5 Deformation of sealant beads in (*a*) butt joints and (*b*) lap joints. (Design by Tore Gjelsvik, Norwegian Building Research Institute)

whereas lap joints subject the scalant primarily to shearing stresses. Figure 2.5 illustrates these two types of movement.

The butt joint is commonly used for exterior panel joints, for expansion joints, and for control joints in masonry walls. The lap joint is used for exterior panel-to-mullion joints, exterior panel-to-sill joints, and the joints between sheet materials.

2.5   Calculation of Joint Movement

In building practice most sealants will be installed in moderate weather: that is, between the extremes of temperature they will experience in service, which means that they will experience deformations on either side of an initial or zero strain condition.

Since butt joints are more common and are generally more critical in terms of the consequences of failure, most computations of joint movement are based on the use of butt joints.

It is quite simple to compute a theoretical value for joint movement, based only on temperature changes. For example, the movement of a $\frac{1}{2}$-in.-wide joint in a 30-ft section of concrete wall (or slab) would be as follows:

Given: Normal size of joint at 70 F: ½ in.
         30-ft.-long concrete panels
         Temperature change: −30 to +130 F
         Coefficient of expansion for concrete: 6.0×10^{-6} in./(in.)
         (deg.)

With the sealant material installed at 70 F, the total joint opening (representing tension in the sealant) is the result of the temperature drop from +70 to −30 F (100 F differential).

Joint movement $= 30$ ft $\times 12$ in./ft $\times (6 \times 10^{-6}) \times 100$ F $= 0.216$ in.

This amount of movement represents a 43% change in the size of the joint. As will be pointed out later, this does not represent the true strain in the sealant material.

This oversimplified calculation is admittedly an approximation because it fails to take into account the many other factors which affect joint movement. However, because of the lack of reliable research data on the actual behavior of buildings, this formula will probably continue in widespread use for a long period of time.

2.6   Stick-Slip Movement

The exterior facade of a structure consists of a number of elements, which may be any of several building materials, each with its own coefficient of expansion. Each of these building parts tends to remain stationary until moved by some outside force. In this case, the force is generally expansion caused by temperature variation. These parts, however, are not subject to uniform movement. As temperature changes, stresses build up within the building part. When these forces become high enough to overcome the inertia of the building part, that part moves. The result is that building joints open and close in a series of short, incremental movements, rather than in a uniform movement cycle.

This stick and slip type of movement is accentuated and aggravated by several factors. The British Building Research Station has calculated the rates of movement of some typical building materials as follows:

Concrete roof coping	0.020 in./hr.
Upper brickwork	0.017 in./hr.
Lower brickwork	0.011 in./hr.
Aluminum mullion	120 in./hr.

The extremely fast response of aluminum to temperature change means that the movement of joints of dissimilar materials, such as the panel-to-mullion joints in exterior walls, depends on the rates of movement of both materials as well as the coefficients of expansion.

Another factor influencing stick and slip movement is the temperature variation between the inside and outside of a building wall. The outside surface of the wall may vary from -20 to $+150$ F; the inside surface of the wall and much of the structural frame to which the wall is attached remains at a relatively constant temperature and humidity. This differential acts to restrain the exterior wall elements, and thereby contributes to nonuniform movement.

2.7   Other Factors Which Affect Joint Movement

The sealants in exterior building joints are considered to expand and contract primarily as a result of temperature change. However, these joints can also move in longitudinal shear due to the slip of a building panel or differential settlement (Fig. 2.6). They may also move in and out in a transverse shearing movement as a result of wind load (Fig. 2.7). The sealant between metal building panels may be flexed about its longitudinal axis as the panel edges move due to temperature changes (the "oilcan effect"). (See Fig. 2.8).

The problem is compounded by other effects. The method of attachment of a curtain wall panel to the structural frame exerts a restraining effect on movement. The color of the panel, moisture absorption, compass orientation, the amount of shade on the building, height and mass of the structure, and wind load all play an important part in the movement of building joints. Some excellent beginnings [1] have been made toward incorporating all these effects into a computer program to predict movement, but much remains to be done.

Wind load on a building is generally assumed to cause a more-or-less uniform pressure on the windward side of the building and a combination of a uniform suction and internal pressure on the leeward side. However, recent studies [2–4] of air flow around buildings have shown that projecting mullions, other buildings in the vicinity, roof type, and ground conditions all cause turbulent air flow which can stress a sealant in fatigue.

Another factor to be considered in the movement of exterior joints in a curtain wall is the simple statics of building height. A tall building shaft acts much like a huge beam cantilevered out of the ground. When the top of this beam is deflected by wind load, the joints closest to the

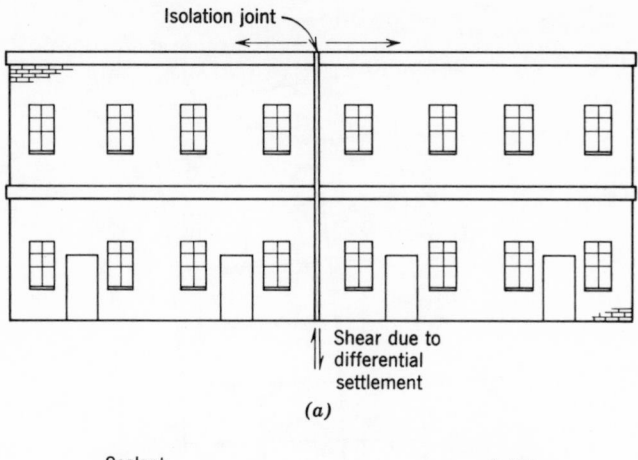

Isolation joint

Shear due to
differential
settlement

(a)

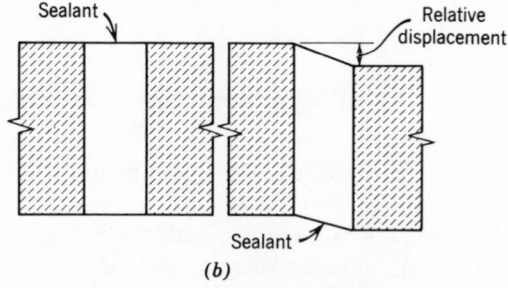

Sealant

Relative
displacement

Sealant

(b)

Fig. 2.6 (a) (b) Longitudinal shear in sealant. Longitudinal shear is a movement of one joint face in a direction parallel to the longitudinal axis of the joint. It could be caused by building settlement or a slight slip of one building panel with respect to the adjacent panel.

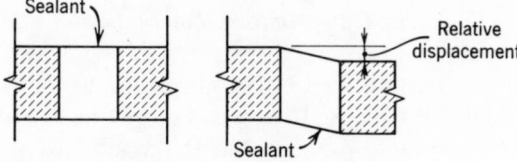

Sealant

Relative
displacement

Sealant

Fig. 2.7 Transverse shear in sealant. This type of strain in a sealant can be caused by one building panel moving with respect to the adjacent panel in a transverse direction across the joint.

Fig. 2.8 Bending (the "oilcan" effect). This is a combination of tension or compression with a displacement of the joint such that the sealant is bent about its longitudinal axis.

Wind
direction

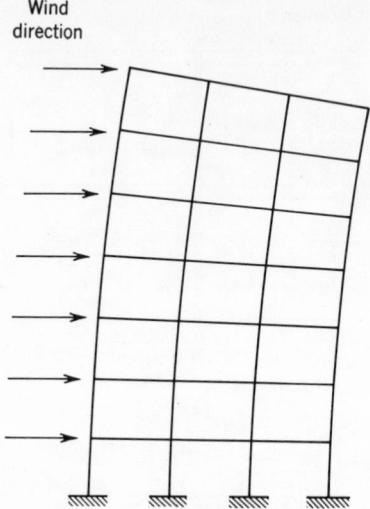

Fig. 2.9 Wind deflection of a building frame.

support (the ground) are naturally subjected to the greatest strain (Fig. 2.9).

In view of all these factors, together with the scarcity of reliable data on building movement, most architects will rely on the approximate temperature formula and professional judgment to provide reasonable limits for joint movement.

2.8 Construction Tolerances

In the erection of a curtain wall building, the joints are usually well detailed on the architect's plans. It is quite simple for the detailer in the office to draw the joints to exact scale. At the job site, it is not so easy for even the most conscientious contractor to conform exactly to the drawings. The joint as it is constructed may differ considerably from the contract plans. The plans may call for ½-in.-wide joints between concrete panels. The joints as constructed may vary from ¼ to ¾ in. in width. The British Building Research Station has published the following data about tolerances in the length and width of large concrete panels.

Note in this table that the total tolerance is not the sum of the individual tolerances. The British Building Research Station has recognized

Type of Production	Site Casting	Normal Factory Casting	Well-Controlled Factory Casting
Tolerance in mold fabrication	± 0.12 in.	± 0.06 in.	± 0.02 in.
Tolerance in slabs cast	± 0.32 in.	± 0.20 in.	± 0.08 in.
Tolerance in erection	± 0.24 in.	± 0.20 in.	± 0.16 in.
Total tolerance	± 0.40 in.	± 0.28 in.	± 0.18 in.

that tolerances may be either plus or minus and has recommended a total tolerance based on measurements of actual panels.

Metal curtain wall panels are factory fabricated under very close control. The total tolerance which might be expected in these panels is less than the tolerance expected in concrete panels. Nevertheless, the metal

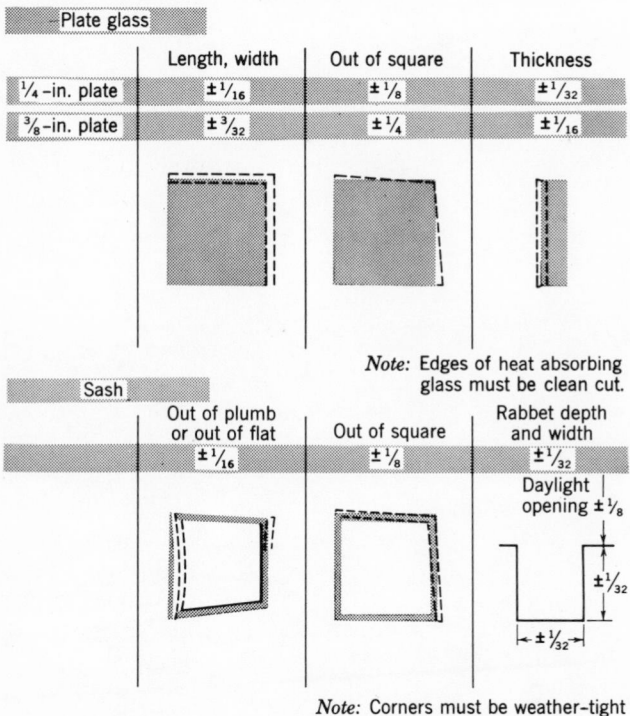

Fig. 2.10 Typical dimensional tolerances. (Courtesy of PPG Industires)

curtain wall building may be expected to have joint sizes which vary ⅛ in. from those shown on the plans.

Glass panes and metal sash may also vary in size. Figure 2.10 shows typical dimensional tolerances for these units.

Why are the tolerances of building parts so important? The joint sealant is one of the last materials to be installed in the building. Consequently, the sealant must fill the gaps between the other building components, whether or not these gaps happen to be the proper size. One quarter of an inch does not seem like much of an error in a large building. But consider a joint in a curtain wall; normal joint width is ½ in. Because of temperature change, the joint is expected to move ¼ in. in extension and ¼ in. in compression, or 50% strain on the sealant in either direction. If the joint as constructed is only ¼ in. wide, the sealant will fail. In extension the sealant moves ¼ in. This is 100% extension of the sealant material which was designed for 50% movement. In compression, if the joint closes a full ¼ in., the joint width is zero and all the sealant is squeezed out of the joint.

3

Stresses and Strains in Sealants

3.1 Strains in Butt Joints

Joint sealing materials play an important part in the overall performance of a completed building and, consequently, should be as carefully
selected and designed as any other building component. Many factors
will ultimately affect the performance of a sealant, but the shape and dimensions of the sealant cross section are considered of primary importance.

The design of the seal, of course, varies with the type of material
being used. Highly elastic mastic-type sealants, deformable mastics,
tapes, gaskets, and foams are different materials and, consequently, have
different criteria for determining optimum performance.

3.1.1 Highly Elastic Sealants

Silicones, urethanes, and the higher modulus polysulfides and polymercaptans may be considered as highly elastic or high recovery sealants. These materials are prime candidates for exterior, one-stage sealing
operations in working joints because they tend to return to the original
shape after the removal of an imposed load.

It is sometimes assumed that the deeper a joint is sealed, the better it
is. However, theoretical calculations [5], laboratory tests, and field experience indicate that this is not the case. There is a definite relationship between the width of joint and the depth of seal that determines
the amount of strain in a sealant. An optimum shape factor has been
found for the highly elastic sealants in butt joints. This optimum seal is

27

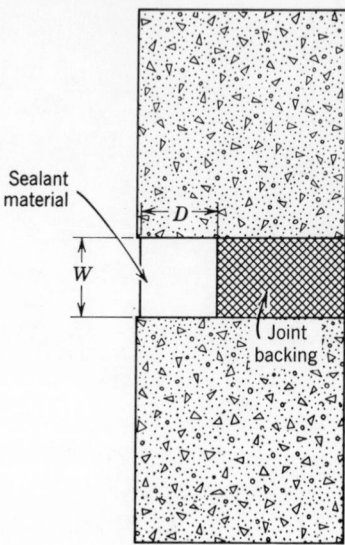

Fig. 3.1 Joint with shape factor of one ($W/D = 1$).

twice as wide as it is deep: that is, a width (W) to depth (D) ratio of two. However, practical considerations usually preclude the possibility of using these optimum proportions. For example, a ⅜-in.-wide joint would be sealed for a depth of only ³⁄₁₆ in. With the tolerances and variations which exist at the job site, this shallow depth of seal would undoubtedly result in a large percentage of failures. Consequently, a width-to-depth ratio of one ($W/D = 1$) appears to be a much more practical joint seal. This shape factor of the sealant material in the joint is controlled by the use of a suitable backing material placed in the joint at the proper depth before sealant installation (Fig. 3.1).

Remember also that the theoretical shape factor derivation deals only with strains in butt joints and assumes a direct proportional relationship between strain and stress. This assumption is approximately true for *only* the high recovery sealants; it does not necessarily hold true for the softer, more deformable mastics.

Quantitatively, the strain in the sealant material may be very simply defined as follows:

$$\text{Strain} = \frac{\text{deformation of the seal}}{\text{initial joint width}}$$

Shown in Fig. 3.2 are two joints with equal initial joint width. Both are extended 100%. The sealant, of course, does not change in volume

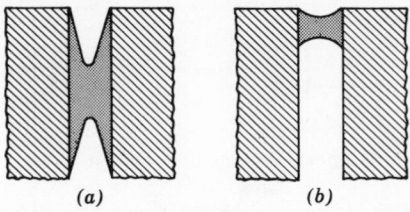

(a)   (b)

Fig. 3.2 Comparison of maximum strains. (*a*) For 2-in.-deep seals. Extension of joint: 100% ; sealant strain: 550% . (*b*) For ½-in.-deep seals. Extension of joint: 100% ; sealant strain: 160% . (Courtesy of Egons Tons)

during extension. Consequently, the deeper seal has to neck down further when both joints are extended the same amount. Note in Fig. 3.2 that the top and bottom surfaces of the sealant material in the deep seal have extended a great deal more than the corresponding surfaces in the shallow seal.

When a bead of elastic sealant material is deformed, the top and bottom surfaces of the sealant neck down into a parabolic shape. Strain can thus be computed by determining the length of this parabolic curve.

Because the top and bottom surfaces of the sealant undergo the greatest change in length (strain), the stresses at these points are naturally maximum. Figure 3.3 shows a qualitative distribution of stresses at the interface. Note that the stresses in the substrate at the edges of the sealant bead become quite high. Note also that the direction of the maximum strain (and, consequently, stress) in the sealant material varies between the deep and shallow seals. In the deep seal, the maximum stress (*P*) direction has a larger vertical component and, consequently, a greater tendency to peel. When this combination of tension and peel stresses exceeds the adhesive strength of the sealant material, a failure is initiated.

Since most sealants are installed in moderate weather, it must be remembered that the sealant bead will be subject to compression as well as tension. The parabolic deformation of the sealant also holds true

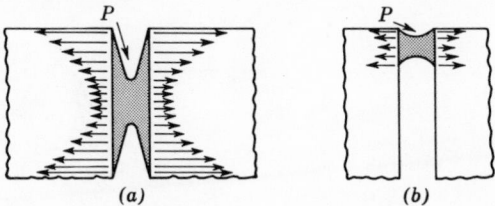

(a)   (b)

Fig. 3.3 Comparison of substrate stresses for (*a*) 2-in.-deep seals and (*b*) ½-in.-deep seals.

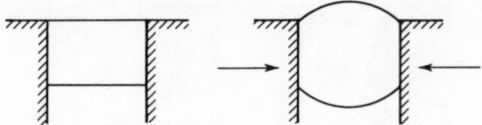

Fig. 3.4 Sealant bulge under compression.

when the sealant is in its compression phase. Figure 3.4 shows that the sealant will extrude outside the joint area under load. This may cause an unsightly appearance and, in traffic-bearing joints such as industrial floors, may be extremely critical in terms of both stress and safety hazard. It is advisable in horizontal joints, such as floors and sidewalks, to use a self-leveling material and to keep the level of the sealant slightly below the floor surface.

An improvement in the rectangular cross section can be achieved by using a curved back-up material and tooling the top surface of the sealant as shown in Fig. 3.5. This tooling of the joint must be done with some care, however, because if the sealant bead is too thin at its center, the sealant under compression will buckle like a slender column. This will cause high peel stresses at the lower corners (Fig. 3.6).

In addition to having the proper shape factor, the elastic sealants must be bonded to only two sides of the joint in order to perform properly. The bottom surface of the sealant must be free to deform. Figure 3.7 shows that if the bottom of the joint is bonded, the sealant must rupture in order to deform. Figure 3.8 shows that the same basic principle of permitting the bottom of the sealant bead to deform applies also to corner joints.

These high recovery sealants have the advantage of returning to the original shape after the removal of an imposed load. This means, then, that these materials remain in a stressed state as long as they are deformed. As a generalization, it is true with rubbery sealants that the higher the recovery, the lower the tear resistance. Consequently, once a

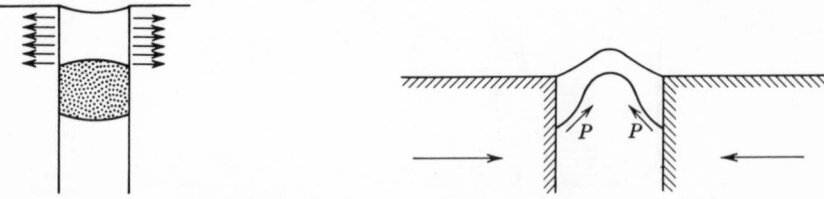

Fig. 3.5 Uniform substrate stress in tooled joint.

Fig. 3.6 Peel stresses in the lower corners of an overtooled joint.

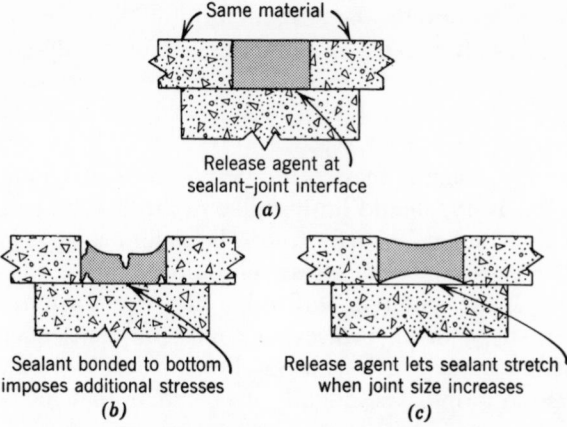

Fig. 3.7 (Courtesy of Machine Design)

failure is initiated in the material, the failure is likely to propagate rapidly, thereby causing extensive damage.

3.1.2   *Deformable Sealants*

The deformable sealants cover a wide range of materials: polysulfides, polymercaptans, butyls, solvent-based acrylics, and "latex" caulks of various composition. The chief characteristics of these sealants as a group are that they show some degree of instantaneous elasticity or "rubberiness" under short-term loads, but will creep or flow under long-term

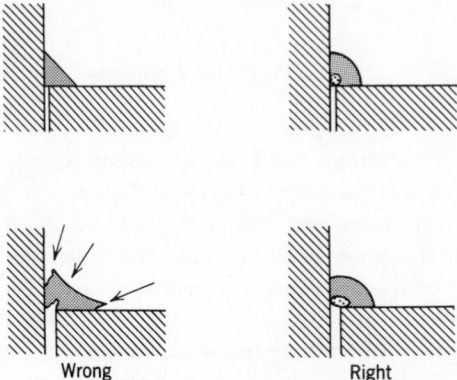

Fig. 3.8 Design of corner joints. (By Tore Gjelsvik; Norwegian Building Research Institute)

loading. Within the group, the sealants exhibit these two properties in varying degrees. Polysulfides, for example, may be highly cross-linked to show a large percentage of instantaneous elasticity (recovery) and only a small amount of cold flow. On the other hand, the solvent-based acrylics will show very little recovery.

It is commonly assumed that all mastic-type sealants, both high recovery and low recovery, should be installed according to the shape factor principle. This is not necessarily true. The deformable or low recovery sealants show a great deal of stress relaxation; that is, when held in a deformed shape, the materials will relax into equilibrium so that they are in a new shape in an unstressed condition. Consequently, with a very deformable sealant such as a butyl or a solvent-based acrylic, it is probably wise to form a seal slightly deeper than that indicated by the optimum shape factor, in order to obtain a greater adhesive area at the interface.

The calculation of stresses in the deformable sealants has been accomplished [6], but the results are of academic value only. What is more important than stress is the amount of shape distortion to which these materials are subject. It is a paradox of joint sealing that these materials, which can generally take more ultimate elongation than the highly elastic sealants, should be used preferably in joints with less movement. Stress is of little consequence since it relieves itself so rapidly. However, if these deformable sealants are subject to excessively large deformations, they tend to wrinkle at the surface and cause an unsightly appearance. The growth of two-stage weatherproofing should lead to increased use of the softer polysulfides, solvent-based acrylics, butyls, and latex caulks because the appearance of the concealed sealant is of such little consequence.

3.1.3 Preformed Shapes

The preformed seals, or gaskets, are generally formed from high recovery elastomers as either rigid extrusions or foams. Because of the principle under which they operate, they depend on high recovery in order to achieve optimum performance. These seals are precompressed and inserted into the joint in the compressed state. As they attempt to return to normal shape they exert a force against the joint wall, thus forming an effective seal (Fig. 3.9). These seals have a unique advantage. Since the cross section of the seal includes a certain amount of void space, these seals can be further compressed without extruding outside the joint area. The compression seals must be carefully designed, however. As the joint opens due to temperature changes, some of the

Fig. 3.9 Preformed seals.

precompression is relieved in the seal. The seal must therefore be exactly sized for its joint opening. If the joint opens too far, the seal ceases to function and may actually slip out of the joint.

Adhesives are used in connection with preformed seals to make them operative under tensile loading. These adhesives have been reasonably successful with some of the preformed foam seals, but not very successful with the more rigid extrusions.

The development of the preformed seals up until this time has been largely empirical. Because of the great variety of complex shapes in use, no thorough stress analysis of these seals had been published. However, the method of photoelasticity is now being used to analyze the preformed seals. Figure 3.10 shows the stress pattern in a typical seal.

3.1.4 Tapes

The tapes may be of either cured (high recovery) or uncured (low recovery) material. The stresses in the tape under load naturally vary depending on the type being used. A compression-deflection test and a stress relaxation test are used in laboratory evaluations of the material, but no rigorous stress analysis has been attempted. Since these materials offer such versatility, more research in this area is certain to come in the future.

3.1.5 Foams

The foam sealants are similar to the rigid gaskets because they contain a large percentage of internal void space. Thus these materials can be subject to compressive loading in service without bulging out of the joint. Adhesives have been moderately successful in helping these materials to sustain low-level tensile loading. As with the more rigid extrusions, their development has been empirical. The only stress data available are the results of experimentation and testing. The critical factor in the performance of a sealing foam is that it must maintain pressure against the joint wall in order to function. Brown [7] has shown that a

Fig. 3.10 Photoelastic stress pattern in a preformed joint seal.

typical neoprene closed cell foam will require a pressure of 15 psi to compress the seal 25%. This value will decay to 5 to 6 psi after one or two years of service. This level of sealing pressure should be sufficient to keep the joint well sealed.

3.2   *Strains in Lap Joints*

Stresses and strains in lap joints have received very little attention. The proper shape factor for use in lap joints has been largely a matter of common sense and judgment, but no derivation has been published. Koppes [8] recommends that the thickness of the sealant in the joint be at least equal to the amount of expected joint movement. Researchers are currently working on lap joints, and therefore a stress distribution and an optimum shape factor should be published shortly.

3.3   *Types of Failures*

Again, the type of failure likely to occur in a sealant depends on the type of sealant being used.

3.3.1   *Mastic Sealants*

The adhesive failure, which is the most common type of sealant failure, is a loss of bond between the sealant material and its substrate (Fig. 3.11).

The cohesive failure is a failure within the body of the sealant material. This failure frequently begins with a small nick or puncture of the sealant (Fig. 3.12).

The spalling failure is not, strictly speaking, a sealant failure, but its effects are just as destructive. If the cohesive strength of the sealant is greater than the cohesive strength of the substrate, this type of failure will occur. The spalling failure is frequently seen with stiff, high recovery sealants when used with low strength concrete panels (Fig. 3.13). Spalling is one of the most frequently encountered failures in highway pavement joints. (see Chapter 19.)

The intrusion failure (Fig. 3.14) is seen in highway pavement joints

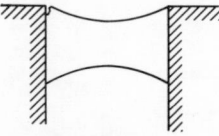   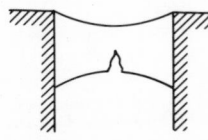   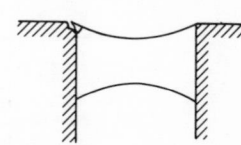

Fig. 3.11 Adhesion failure.   **Fig. 3.12** Cohesion failure.   **Fig. 3.13** Spalling failure.

Fig. 3.14 Intrusion failure.

and other traffic-bearing areas. This failure occurs as a three-step process. The sealant extends, necks down, and then fills with dirt. During
the succeeding compression cycle, this pocket of dirt attempts to close.
This closing action generally abrades the surface of the seal so that it
fails during a subsequent tension cycle.

Two additional types of failure are common, both of which are pecul-
iar to the deformable sealants. One type occurs when tension is the first
strain applied to the sealant; the other type occurs when compression is
the first strain to be applied. The tensile failure occurs as follows. The
material is extended and held in a necked-down shape (Fig. 3.15). The
sealant relaxes into equilibrium in this new shape. When the joint closes, the sealant acts like a column and buckles. In addition to exposing
the strained sealant to further weathering, this may cause high peel
stresses at the lower corners. The compression failure is shown also in
Fig. 3.15. The sealant is compressed and relaxes into equilibrium in this
new shape. When the joint opens, the sealant does not return to its
former rectangular shape; instead, it yields at its minimum cross section,
which is immediately adjacent to the joint interface. This high concen-
trated strain may lead to immediate failure.

Remember also that the stress and strain computations which have
been made concerning mastic sealants all *assume* that the sealant is

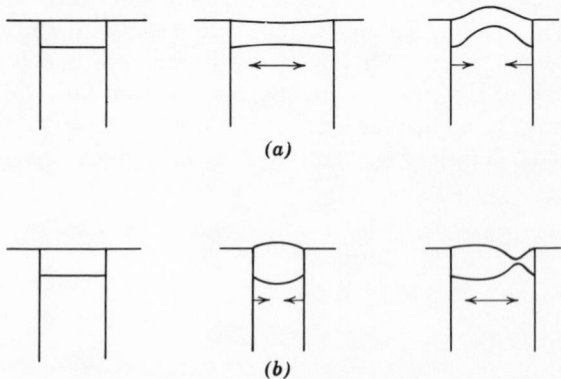

(a)

(b)

Fig. 3.15 Change in sealant shape due to flow. (a) Viscous tension–compression effect.
(b) Viscous compression–tension effect.

fully cured before it goes to work. This assumption is far from true. The building, bridge, or pavement is not so obliging as to wait until the sealant cures before it begins to expand and contract. The structure is constantly moving, and the sealant must be expected to function as soon as it is placed in the joint. The effect of this joint movement on the uncured sealant has never been investigated and is not considered in any of the standard specifications.

3.3.2 Preformed Seals

The most common failure of the preformed seals is a loss of interface pressure between the seal and the substrate. The usual cause of this failure is simply poor design, which results in the wrong size seal being placed in the joint. In order to perform properly, the seal must continue to exert pressure against the interface when the joint is at its widest. On the other hand, when the seal is compressed the voids may be closed, but the seal should not be compressed so far that the rubber itself is stressed in compression.

The preformed seals are generally fabricated of high quality elastomers; hence failures within the seal itself are rare. Some instances have been noted of extrusions cracking after weather exposure, but these have generally been traced to poor rubber compounding.

The preformed seal, however, depends on straight uniform joint walls in order to function. Consequently, when the extrusions are used in conjunction with metal curtain wall panels, any warping or nonuniform distortion of the panel causes the seal to lose its effectiveness. When they are used with concrete panels or slabs, as in highways, industrial floors, or terraces, any spalling of the concrete at the joint causes the seal to be ineffective.

The foams which are used as seals are generally of closed cell urethane or neoprene. These materials have the advantage of being less rigid than the extrusions, so that straight joint walls are not so critical. In addition, there is little or no lateral spread to these materials when compressed and therefore they do not extrude out of the joint. However, the foams are subject to a slight amount of flow and stress relaxation, which may lead to some loss of seal effectiveness over a long period of time.

3.4 Stress Determination by Experimental Methods

The actual determination of the stress distribution within the body of a sealant material is difficult to establish analytically. When dealing with

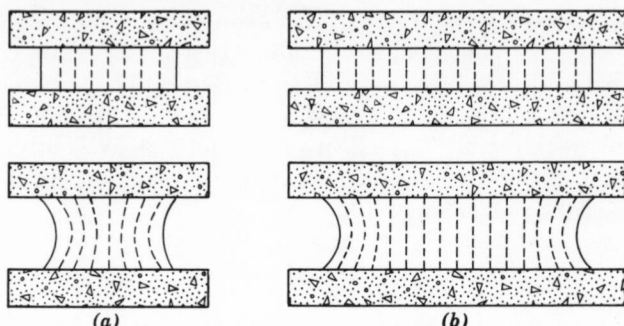

Fig. 3.16 (*a*) 2-in.-long and (*b*) 4-in.-long specimens, both extended 100%. Dashed lines marked on the specimens clearly indicate the directions of the strains and stresses in the sealant.

crystalline materials which undergo small deformations, it is relatively simple to use strain gauges, and thus determine stress experimentally from the modulus relationship. With the rubbery materials, however, the deformations are large and the stress-strain relationship is nonlinear, so that analytical methods become complicated and tedious.

Visual methods of strain determination can be used very effectively with sealant materials. One quite simple method of determining a qualitative strain distribution requires only a testing machine and a felt tip marking pen. Lines are drawn on the specimen at equal intervals. When the specimen is extended, the directions of the strains are immediately apparent. Figure 3.16 shows the differences in the strain directions in two sealant specimens.

Quantitatively, the method of photoelasticity offers the best solution for the stresses in a sealant. With this experimental method, it is possi-

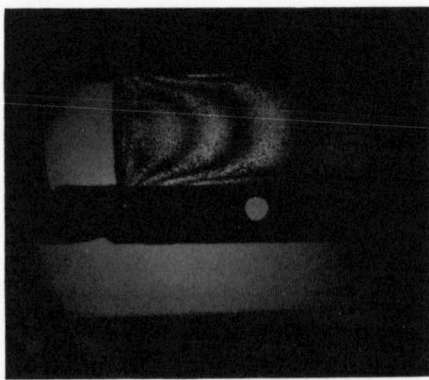

Fig. 3.17 Photoelastic stress patterns in a sealant specimen under compressive load.

ble to literally look inside the specimen and see the stress distribution. Certain translucent materials, such as the urethanes and polysulfide epoxy combinations, are artificially birefringent materials. This means that these materials will show a pattern of stress lines when loaded and placed in a polariscope. Thus specimens in the proper size and shape can be cast of this translucent material, and the stresses can be determined experimentally. Correlation between the translucent models and the opaque sealant material is then established by plotting modulus curves for the two materials. Figure 3.17 shows the stress pattern in a loaded sealant specimen.

The author has successfully used this method to verify Tons' original work on shape factors: to determine the stress relaxation rate of polysulfide sealants, to determine the stresses in preformed highway seals, and to study the length of sealant specimens. Research is currently under way to determine the optimum shape factor for shear stresses in lap joints and also to determine the maximum stresses in structural glazing gaskets.

3.5 *Fatigue*

Fatigue in a construction material is essentially failure caused by a large number of repetitions of stress. In mastic sealants, fatigue is a complicated phenomenon involving not only stress and strain, but also the molecular structure of the polymer. Some experimental work has been done on the "work hardening" of polysulfides and urethanes, but no complete fatigue study has ever been published.

4

Properties of Sealant Materials

The user of any sealant product, whether he be caulking contractor, homeowner, maintenance engineer, or architect, is really looking for only one property in a sealant. What he wants to know is, "Can I put it in the joint and forget it?" or "Will it seal the joint effectively and keep it sealed for a long period of time?"

The manufacturer, for his part, is also earnestly seeking the answer to the same question. However, the manufacturer must be able to put the answer to this question not only in quantitative terms, but also in terms that the consumer can understand and use.

There is no such thing as the perfect or ideal sealant. No single sealant is the panacea that will solve every problem. Different joint shapes and conditions of movement demand different sealants, so the manufacturers have had to develop a wide range of tests to cover all possible conditions.

A partial list of some of the chemical and physical tests used to evaluate sealants is shown below:

Solvent resistance	Ultimate elongation
Toxicity	Creep
Dielectric strength	Stress relaxation
Aging	Peel strength
Weathering	UV resistance
Ozone resistance	Adhesive strength
Hardness	Tear resistance
Modulus of elasticity	Cyclic tension and compression

All these tests are valid and generally defined in terms of either some ASTM test requirement or the requirements of one of the standard specifications. Out of this maze of properties and tests, each defined in its particular units, it is very difficult for even the experienced architect or engineer to know exactly what he is buying.

An attempt will be made here to define some of the more important properties of sealants and methods of testing, in order to facilitate the selection of the proper sealant for specific applications. The limitations of some of these tests will be pointed out. Quantitative values for the various tests as they apply to specific sealants will be outlined in the chapters covering the respective sealants.

4.1 Toxicity

Strictly speaking, the toxicity of sealant materials is not at question. Since many of the modern sealants are sophisticated chemical compounds and since many contain solvents, these materials may be allergents. Continuous exposure may cause some skin or eye irritation. Consequently, as with most chemicals, it is advisable to use caution and avoid repeated or prolonged contact with the skin. Since allergy reaction varies so widely from person to person, there is no real test to define this property.

4.2 Solvent Resistance

Most manufacturers test the resistance of their own products to a wide range of solvents, including oils, alcohols, and mild acids. In quantitative terms, this resistance is generally determined by the requirements of the test procedure outlined in ASTM D-471. This is a test of oil resistance. There is also a requirement in one of the Federal specifications used for airport paving work, which requires resistance to the spillage of jet fuel.

4.3 Aging and Weathering

The aging or weathering of sealants in service generally shows up as a crazing or "alligator cracking" of the sealant surface. Weathering may be the combined effect of solvent evaporation, ozone attack, migration of plasticizers, UV attack, water immersion, and other factors.

Fig. 4.1 An Atlas Weatherometer.

Potentially, of course, any crack in the surface of a sealant in a working joint is a trouble spot. There are indications, however, that crazing may be largely a surface phenomenon which affects the appearance of the joint more than anything else.

The accelerated aging and weathering of sealants by a weatherometer device is described in several ASTM test procedures. The specimens are exposed in a weathering chamber which subjects them to alternating cycles of UV light and water spray. Although more correlation data is badly needed, if a material withstands 1000 hr or more in the weatherometer, it is usually considered adequate for exterior joints (Fig. 4.1).

Fig. 4.2 A Shore Durometer.

4.4 UV Resistance

The resistance of a sealant to deterioration by ultraviolet rays may be very important in glazing work. The adhesive bond of a sealant to glass may be affected by light penetrating the glass. USASI Specification 116.1 includes a tensile adhesion test after UV exposure. This is an excellent specification requirement, but it might be strengthened by adding a peel adhesion requirement.

4.5 Hardness

The hardness of a sealant material is not in itself an indication of how the material will perform in a building joint. However, since the hardness is a resistance to penetration, it is sometimes used as a quick measure of the modulus of elasticity. Hardness is measured by a Shore Durometer (Fig. 4.2); the scale numbers from 0 to 100 are strictly relative. For example, on the Shore A range an automobile tire has a hardness of about 70, an automobile inner tube has a hardness of about 40, and a pink rubber eraser has a hardness of about 15. The hardness of mastic sealants for buildings should generally be in the 10–30 Shore A range. Sealants softer than this are easily gouged from building joints, and in pavement joints they may be penetrated by stones or spike heels. Sealants harder than 40 are generally too stiff to have sufficient compliance to accommodate building movement. Neoprene gaskets, however, are generally fabricated of rubber in the 40–70 Shore A range because they operate only in compression and thus may contain internal void space.

The hardness of most sealants varies with temperature. These materials soften with heat and stiffen at colder temperatures. Quantitatively, many sealants will increase in hardness by 10 to 20 points in very cold weather. That is, a material with a Shore hardness of 20 at 70 F may

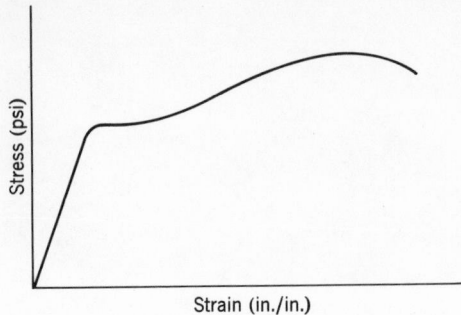

Fig. 4.3 Stress-strain curve for low carbon steel.

have a hardness value of 35–40 at −20 F. This increase should be kept in mind when selecting a sealant material.

4.6 Modulus of Elasticity

The modulus of elasticity of a sealant material is one of the most informative properties of the material to the laboratory technician, but at the present time it is of little value to the specifier in the selection of a sealant. If the testing procedure could be standardized, much valuable information would be conveyed to the consumer.

The modulus of elasticity of any material, or the ratio of stress to strain, is determined experimentally as the slope of the stress-strain curve at some point. Most materials, such as the metals (Fig. 4.3), have a stress-strain curve which includes an initial straight-line portion and therefore the selection of a modulus value is relatively simple. However, the rubbery materials have a continuously curving stress-strain plot, and hence the modulus slope in this case must be arbitrarily defined. At the present time, many laboratories will plot the stress-strain curve and draw a secant to the curve at some value of strain. The slope of this secant line defines the modulus. However, different laboratories use different strain values so that the modulus value is hard for the consumer to interpret. Modulus values of 30, 50, 100, and even 300% strain have been reported in the literature. Figure 4.4 shows a stress-strain curve with a 50% modulus value reported.

The modulus of elasticity is affected by both specimen shape and testing rate. A flat specimen cut out of a pressed sheet is often used to determine the modulus. This is probably the best available shape for determining the properties of the rubber to be used in a preformed gasket, but is quite unrealistic in evaluating a mastic sealant. The mastic seal-

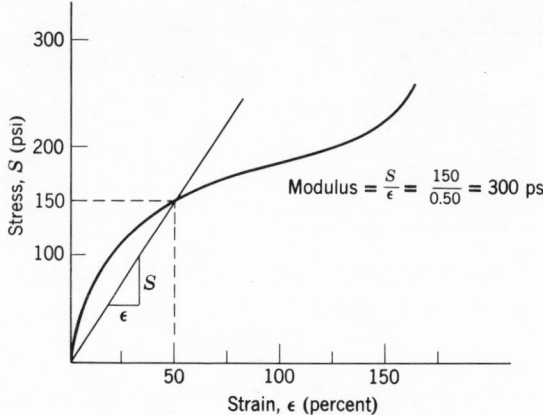

Fig. 4.4 Typical stress-strain curve for an elastomeric sealant.

ant should be formed into a ½ × ½-in. specimen which is more representative of the actual shape of the sealant bead in the joint. Also, the faster the testing rate, the higher the modulus of elasticity.

4.7   *Ultimate Elongation*

The ultimate elongation of a sealant is one of the properties usually shown on the manufacturers' technical data sheets. It is considered to be a measure of the stretchability of the sealant. However, the values shown on the "tech data" sheets generally have little or no relationship to the amount of extension the sealant can endure under service conditions. For example, it is not unusual to see "Sealant A" advertised with an ultimate elongation value of 500%. This is not misrepresentation on the part of the manufacturers. For, in order to obtain reproducibility of results, the manufacturers have had to rely on standardized tests based on ASTM procedures. Specimen shape is again the culprit. These tests are generally conducted with thin, flat tensile specimens which are quite stretchy. The same "Sealant A" formed into the USASI specimen, which is ½ × ½ × 2 in., would probably not extend over 50%.

It is possible, of course, to compound sealants which can be extended as much as 1000%, but they might still make very poor sealants. The recovery of the sealant must be considered; more importantly, especially with concrete substrates, the load-deformation (stress-strain) relationship must be considered. It matters little that a sealant can be extended 500% if the load required to produce this deformation is so high that it will deform or fracture the substrate. This load-deformation relationship

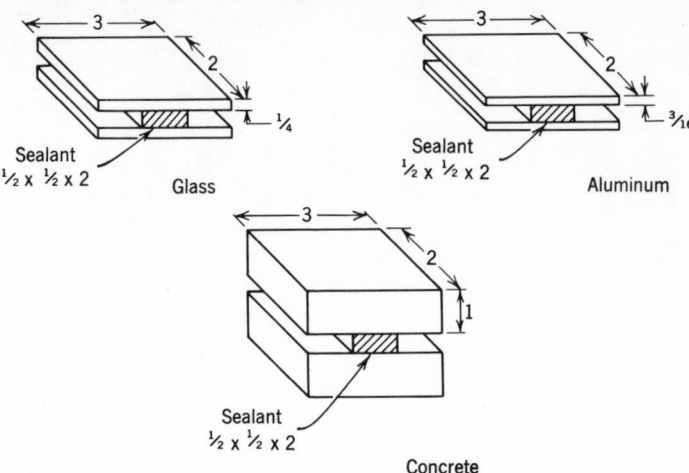

Fig. 4.5 Typical bond specimens.

is precisely what is furnished by the stress-strain curve, again emphasizing the importance of the modulus of elasticity.

4.8   Adhesive Strength

The adhesive strength of a sealant may be measured by a tensile adhesion (bond-extension) test, a cyclic tension and compression test, or a peel adhesion test. The USASI specification 116.1 contains bond-extension tests for initial adhesive strength, adhesive strength after heat aging, adhesive strength after water immersion, and adhesive strength after UV exposure. Figure 4.5 shows typical bond test specimens. This specification also requires a cyclic test. Federal specification TT-S-00230 contains a bond-extension test and a peel adhesion test. These requirements have served the purpose of providing some measure of order to the testing of sealants, and they have brought the sealants industry a long way toward maturity. However, these requirements are now in need of further updating.

For example, the "cyclic" tests contained in most of these specifications require that the specimens be extended and then be allowed to recover or else be compressed back to the original width. No actual compression cycle is required, in spite of the fact that nearly half of all building movement will stress the butt joint in compression. The lone exception is the new TT-S-00230b specification which now requires a compression phase.

The criteria for determining failure as spelled out in these specifications are vague and hard for the potential user to evaluate. Failure is defined as an adhesive or cohesive separation exceeding $\frac{1}{10}$ in. in depth. The potential user knows only that the material "passed" this requirement; he has no way of knowing how the material actually survived the test. For example, the material may develop a slight tear or loss of adhesion and still pass the test. However, most rubbery materials, if notched or punctured, will fail completely within the next one or two extension cycles.

The peel adhesion test, when used in addition to the tensile adhesion test, gives a good indication of how a sealant might perform in service. Lap joints which undergo a shear-type movement exert a strong peeling action on the bonded surface of the sealant. None of the standard specifications includes a shear test. Although the peel test is not strictly a shear test, it does give some indication of the adhesive strength of a sealant in shear.

Specimens for the test are usually 1 in. wide and about 6 in. long. The specimens are formed between a metal strip and a strip of heavy cotton duck. The cloth is rolled into and impregnated with the sealant. After the sealant has cured, the cloth strip is folded back 180 degrees and the specimen is pulled apart. The load required to pull the sealant from the metal substrate is reported in pounds per inch of width. Peel adhesion values of 15 to 20 lb/in. indicate good peel strength in a mastic sealant.

4.9 Cyclic Tension and Compression

In spite of there being no specification requirement for a cyclic tension and compression test, many manufacturers and other agencies have incorporated this test into their own evaluation process. Listed below are a few of the testing devices which have been developed. Although this list does not attempt to delineate all the specialized pieces of equipment in use, it is broadly representative of the types of equipment which have been developed.

4.9.1 Bostic Tester

This testing machine, built by Bostic, Ltd. of England, is capable of testing both butt and lap joints because it incorporates both a tension and a shear set-up. The apparatus subjects the specimens to incremental movements which are controlled by time-limit switches. The capacity of the machine is six 2-in.-long specimens at a time on each unit; and the machine is equipped with three units, two for tension-compression cycling and one for shear. The apparatus can be enclosed in an environmental chamber.

4.9.2   TNO Tester

This apparatus was developed by W. VanAs at the Institute TNO for Building Materials in the Netherlands. The machine tests one specimen at a time in a tension-compression cycle. Specimen size is 10 mm × 10 mm × 30 cm (roughly ½ × ½ × 12 in.). This apparatus includes a dynamometer for measuring load and an X-Y plotter for construction of a cyclic stress-strain diagram. The apparatus includes an environmental chamber.

4.9.3   Dominion Tester

This apparatus was developed by the Dominion Rubber Company of Canada. A compression-extension tester, it has a capacity of 24 samples, each 6 in. long. This apparatus, which includes a temperature chamber, applies an incremental movement to the specimens. The time switches give this machine great flexibility in programming the test cycle.

4.9.4   Sika Tester

Developed by the Sika Chemical Company, this machine tests six 2- or 3-in.-long specimens in a tension-compression cycle. Wheel-mounted, it can be moved into a cold room for testing. The machine is built entirely of aluminum and stainless steel to prevent corrosion from condensation.

4.9.5   PCA Tester

This apparatus was developed by the laboratories of the Portland Cement Association. It is the only American-built machine to include load cells for measuring stress as well as strain. The apparatus tests 20 4-in.-long specimens.

4.9.6   ORD Tester

The testing machine developed by the Ohio River Division Laboratories of the Corps of Engineers is probably the largest sealant-testing apparatus in existence. This machine will test 120 2-in.-long specimens in a controlled environment.

4.9.7 Bureau of Reclamation Joint Simulator

This apparatus was designed for testing canal sealants. The apparatus will test two 1-ft-long specimens which are partly immersed in water. This is an outdoor testing device, and the joint movement is actuated by black plastic rods which are affected by temperature changes.

Since only one specification actually requires a true cyclic test, this testing is done for the "in house" evaluation of materials. A quick glance at the testers described reveals a great deal of disparity in the size and shape of specimens which are tested. Each laboratory has an investment in equipment and a familiarity with its own procedures, hence there is a reluctance to change. It is hardly likely that this valuable test will be standardized at any time in the near future.

4.10 Creep, Stress Relaxation, and Recovery

The problems of creep and stress relaxation are intimately related. Both are time-dependent measures of the amount of flow in a sealant under load.

Creep = deformation with respect to time under constant load
Stress relaxation = load with respect to time under constant deformation

Stress relaxation is sometimes described by the somewhat limiting term "compression set," and the recovery of sealants is generally reported at the end of a compression set test. Specimens are blocked in place at a fixed deformation and allowed to remain for a period of time. At the end of this period the specimens are freed and the percent of recovery is measured. Figure 4.6 shows a set of Recovery Test specimens.

The sealant in the joint is actually in the stress relaxation situation; that is, the deformation is imposed on the sealant. The joint moves and the sealant must accommodate itself to this movement. When the sealant is held in the deformed position, the material begins to flow and readjust itself. All sealants will exhibit this phenomenon to some degree.

Stress relaxation or compression set is especially important to the preformed gasket seals. This type of seal is compressed and placed in the joint in the compressed state. Consequently, if there is any appreciable

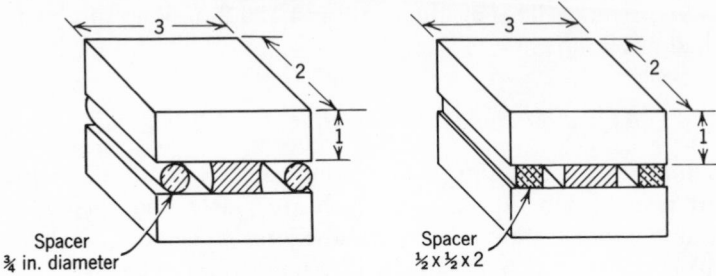

Fig. 4.6 Recovery Test specimens.

amount of stress relaxation, the seal becomes ineffective.

Silicone sealants have very little stress relaxation and, consequently, will show almost complete recovery after the removal of an imposed load. At the other end of the spectrum, the solvent-based acrylics and butyl caulks show very little recovery.

Paradoxically, although the sealant in the building joint is in the stress relaxation situation, the creep property is much easier to measure. Figure 4.7 shows a typical creep test in progress. The creep of a sealant is generally reported as a curve of deformation versus time under constant load (Fig. 4.8). The creep curve is valuable to the formulator because it indicates three distinct phases of sealant behavior. The behavior of the material may be divided into (1) instantaneous elasticity, (2) delayed elasticity, and (3) flow.

The stress relaxation of a material may be reported as a curve of stress versus time at constant deformation; or as a "stress relaxation

Fig. 4.7 Creep test of three sealant specimens.

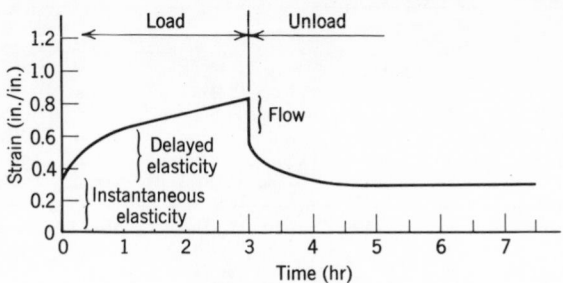

Fig. 4.8 Creep curve.

time," which is the time required for the stress in the specimen to decay to 36.8% of its initial value (Fig. 4.9). The stress relaxation is conducted by extending the sealant specimen 25 to 50% in a tensile testing machine. The specimen is extended as fast as the testing machine will operate. The machine is then locked into position with the sealant specimen at a constant deformation. Then the decay of stress with time (as the material relaxes) is plotted.

In the final analysis, what the consumer wants to know is how these properties affect the performance of the sealant in the joint. Unfortunately, neither the creep nor stress relaxation test lends itself to data which the potential consumer can readily use. Consequently, the recovery of the sealant is the property generally reported. Recovery is the only measure of flow characteristics which is called for in any of the standard specifications.

The inference in the recovery test is that only a high recovery material makes a good sealant. However, there are many instances, such as heel beads for glazing and joints with irregular shapes, in which the low recovery sealant is the better choice. The air seal in two-stage weatherproofing frequently indicates the low recovery sealant as the better

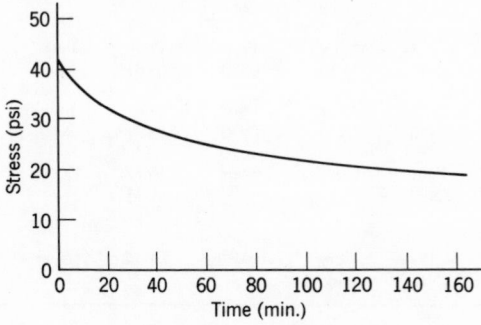

Fig. 4.9 Stress relation curve.

Table 4.1 Nontraffic-Bearing Sealants

Properties	Flexible Epoxy Crack Filler	Drying Oil-Based Caulk	Nondry Poly-butene Caulk or Tape	Non-cure Polyiso-butylene bed Compound	Curing Polyiso-butylene or Butyl Sealant	1-Part Nitrile or Butyl Seam Sealant	1-Part Neo-prene Sealant	1-Part Hypa-lon Sealant	1-P Pol acry Seal
Tack free time @75 F 250° / relative humidity	2 hr	1–3 hr	° ° ° °(NR)	(NR)	2–4 hr	10–20 min	15–30 min	12–48 hr	2–8
Cure time (days) $\frac{1}{4} \times \frac{1}{4}$ in. Channel	3	120	(NR)	(NR)	120	7	30	180	3(
Tensile strength (psi)	1200–1500	2–4	0–1	4–10	10–20	250–400	100–200	20–40	40–
Elongation (%)	15	5	(NR)	20	25	75–125	20–40	15–50	25–
Modulus @ 100% elongation	(NR)	(NR)	(NR)	(NR)	(NR)	200–300	(NR)	(NR)	(N)
Shore A initial	75	(NR)	(NR)	(NR)	10–18	40–60	50–65	12–25	20–
Recovery after 100% elongation	10	(NR)	(NR)	(NR)	15	35	22	20	2(
Continuous service range (deg. F)	−30 +300	−10 +180	−60 +150	−45 +210	−25 +210	−40 +268	−25 +210	−25 +225	− +2
Water immersion properties	Excell.	Very poor	Poor	Fair	Fair	Good	Fair	Fair	Fa
Primer recommended underwater	None	(NR)	(NR)	° R or M	° R or M	None	° R or M	° R or M	° °
Dilute acid resistance	Excell.	Fair	Good	Good	Very good	Very good	Good	Fair	Fa
Dilute alkali resistance	Excell.	Very poor	Very poor	Very poor	Fair	Good	Fair	Good	Fa
Solvent resistance	Excell.	V. poor (NR)	V. poor (NR)	Poor (NR)	Poor (NR)	Very good	Good	Fair (NR)	Goo
Fire resistance	Good	Poor (NR)	Poor (NR)	Poor (NR)	Poor (NR)	Fair	Good	Very good	Fa (N)
Electrical insulation properties	Very good	Poor (NR)	Good	Good	Very good	Good	Fair (NR)	Poor (NR)	Goo
Aging properties	Very good	V. Poor (NR)	Very good	Excell.	Good	Good	Fair (NR)	Good	Ve goo

° Chlorinated rubber.
° ° Moisture-cured polyurethane.
° ° ° Aminosilicone.
° ° ° ° Not recommended for this use.

Reprinted by courtesy of Adhesive A

Table 4.1 (*continued*)

1-Part Poly-sulfide	1-Part Silicone	1-Part Poly-ether Urethane Sealant Shore A less than 25	2-Part Urethane Shore A 20–35	2-Part Ure-thane Shore A 40–60	2-Part Poly-sulfide Shore A less than 25	2-Part Poly-sulfide Shore A greater than 25	2-Part Poly-sulfide Epoxy Sealant	2-Part Poly-sulfide Coal Tar Sealant	2-Part Urethane Coal Tar Sealant
2–48 hr	2–4 hr	12–24 hr	8–24 hr	8–24 hr	6–12 hr	4–12 hr	2–8 hr	1–4 hr	12–24 hr
60	5	30	3–6	3–4	4–7	4–7	3–7	3–4	7–10
00–200	60–150	250–500	150–400	250–600	75–200	100–300	800–1200	60–125	100–300
50–200	100–200	300–450	250–600	200–350	200–350	150–250	40–50	75–150	100–150
75–100	40–80	30–80	50–100	100–150	40–100	75–100	(NR)	5–15	5–15
30–45	25–35	30–50	20–35	40–60	15–24	25–55	65–80	35–50	40–60
45	83	90	60–75	75–82	33–65	40–65	25	50	70
−45 +250	−90 +400	−60 +275	−60 +275	−45 +275	−60 +250	−45 +265	−20 +225	−25 +210	−30 +210
Good	Good to fair	Fair to good	Very good	Very good	Good	Very good	Good	Fair	Excell.
° M or °° V	°°° S	° M or °°° S	°°° M or °°° S	°° M or °°° S	°° M or °°° S	°° M or °°° S	None	° R or °° M	°° M or ° R
Very good	Good	Good	Very good	Very good	Very good	Excell.	Very good	Good	Excell.
Good	Fair	Very good	Excell.	Excell.	Excell.	Excell.	Excell.	Very good	Excell.
Excell.	Excell.	Excell.	Excell.	Excell.	Excell.	Excell.	Excell.	Very good	Very good
Good	Excell.	Very good	Very good	Very good	Very good	Very good	Fair (NR)	Fair (NR)	Excell.
Good	Very good	Excell.	Excell.	Excell.	Very good	Excell.	Very good	Good	Good
Very good	Excell.	Excell.	Excell.	Excell.	Excell.	Excell.	Very good	Good	Good

Table 4.2 Average Life Expectancy (Years) of Nontraffic-Bearing Sealants

Joint Size and Performance Conditions	Surface	Flexible Epoxy Crack Filler	Drying Oil-Based Caulk	Nondry Poly-butene Caulk or Tape	1-Part Noncure Polyiso-butylene bed Compound	1-Part Curing Polyiso-butylene or Butyl Sealant	1-Part Nitrile or Butyl Seam Sealant	1-Part Neo-prene Sealant	1-Part Hypa-lon Seala
Hairline cracks—to 1/8 in. wide joint above grade	Concrete	11	0 (NR)	(NR)	(NR)	(NR)	7	4	(NR)
	Wood	9	1 (NR)	(NR)	(NR)	(NR)	6	5	(NR)
	Metal	10	0 (NR)	(NR)	(NR)	(NR)	8	5	(NR)
	Glass	10	0 (NR)	(NR)	(NR)	(NR)	5	3	5
1/8 to 1/4 in. wide above grade	Concrete	12	(NR)	(NR)	B	5	5	6	6
	Wood	10	2 (NR)	(NR)	B	4	6	5	7
	Metal	12	(NR)	B	B	5	8	5	7
	Glass	10	2 (NR)	B	B	6	8	5	6
1/8 to 1/4 in. wide under water	Concrete	12	(NR)	(NR)	B–M	3	1 (NR)	5–R	5
	Wood	(NR)	(NR)	(NR)	(NR)	(NR)	(NR)	(NR)	(NR)
	Metal	(10)	(NR)	(NR)	S–M	2 (NR)	3	3–V	4
	Glass	8	(NR)	(NR)	B	2 (NR)	2 (NR)	2–S (NR)	3
	Plastic	(NR)	(NR)	(NR)	B	3	5	5	4
1/4 to 3/4 in. wide above grade	Concrete	(NR)	(NR)	B-M	B	5	(NR)	5	6
	Wood	(NR)	2 (NR)	B-M	B	3	2 (NR)	5	5
	Metal	(NR)	(NR)	B	B	4	2 (NR)	5	5
	Glass	(NR)	2 (NR)	B	B	5	3	6	4
	Plastic	(NR)	(NR)	(NR)	B	3	5	5	4
1/4 to 3/4 in. wide under water	Concrete	(NR)	(NR)		B–M	2 (NR)	(NR)	5–M	(NR)
	Wood	(NR)	(NR)		(NR)	(NR)	(NR)	(NR)	(NR)
	Metal	(NR)	(NR)		B–V	1 (NR)	(NR)	3–V	(NR)
	Glass	(NR)			B–S	2 (NR)	(NR)	1–S (NR)	(NR)
	Plastic	(NR)		Not to be used	B–V	3	(NR)	5–V	(NR)
3/4 to 1 1/2 in. wide above grade	Concrete	(NR)		under	B	(NR)	(NR)	(NR)	(NR)
	Wood	(NR)	NR	poly-	B	(NR)	(NR)	(NR)	(NR)
	Metal	(NR)	except	sulfide	B	(NR)	(NR)	(NR)	(NR)
	Plastic	(NR)	for tem-	poly-	B	(NR)	(NR)	(NR)	(NR)
	Glass	(NR)	porary	urethane	B	(NR)	(NR)	(NR)	(NR)
3/4 to 1 1/2 in. wide under water	Concrete	(NR)	use, and	silicone	B–M	(NR)	(NR)	(NR)	(NR)
	Wood	(NR)	then on	sealants	(NR)	(NR)	(NR)	(NR)	(NR)
	Metal	(NR)	non-		B–S and V	(NR)	(NR)	(NR)	(NR)
	Glass	(NR)	porous		B–S	(NR)	(NR)	(NR)	(NR)
	Plastic	(NR)	surfaces		B–S	(NR) *	(NR)	(NR)	(NR)

* The Butyl sealant must be of the type that will not harden beyond a Shore A of 40.

Reprinted by courtesy of *Adhesive Age*.

Explanation of Numbers and Letters for Tables 4.1 and 4.2

B = Backing or bedding filler only   R = Chlorinated rubber primer
(NR) = Not recommended for this use   S = Aminosilicone primer
P = Pourable (self-leveling) grade   V = Vinyl or butyral solution primer
M = Polyurethane primer

Table 4.2 (continued)

1-Part Poly-acrylic Sealant	1-Part Poly-sulfide	1-Part Silicone	1-Part Polyether Urethane Sealant Shore A less than 25	2-Part Urethane Shore A 20–35	2-Part Poly-sulfide Shore A less than 25	2-Part Poly-sulfide Coal Tar Sealant	2-Part Urethane Coal Tar Sealant	2-Part Silicone Sealant	1-Part Vinyl-acrylic Cork-Filled Sealant
(NR)	(NR)	(NR)	(NR)	5–P	(NR)	(NR)	(NR)	(NR)	(NR)
(NR)	(NR)	(NR)	(NR)	8–P	(NR)	(NR)	(NR)	(NR)	(NR)
(NR)	(NR)	(NR)	(NR)	4–P	(NR)	(NR)	(NR)	(NR)	(NR)
5	7	10	(NR)	2–P–S	(NR)	(NR)	(NR)	(NR)	(NR)
5	8–R	7–M	9–M	10–M	8	(NR)	(NR)	8	5
7	8–M	8–M	10–M	10	9	(NR)	(NR)	(NR)	4
7	11	10–S	8–S	8–S	10	(NR)	(NR)	(NR)	4
7	9	10–S	8–S	10–S	9	(NR)	(NR)	10–S	5
5–V	6–M	5–M	6–M	9–M	7–V	(NR)	(NR)	(NR)	(NR)
(NR)	(NR)	(NR)	(NR)	(NR)	(NR)	(NR)	(NR)	(NR)	(NR)
4–V	6–S	6–S	5–V	4–S	5	(NR)	(NR)	(NR)	(NR)
4–S	4–S	7–S	5–S	5–S	3–S	(NR)	(NR)	8–S	(NR)
5	2 (NR)	6–S	5–S	6–S	4–S	(NR)	(NR)	(NR)	3
5	8–M	7–M	9–M	8–M	8–R	7	8–M	(NR)	4
4	6–M	8–M	10–M	10–M	9–R	8–M	9–M	(NR)	3
4	10	9–S	7–S	10	11	5–V	7–S	(NR)	5
4	11	10–S	9–S	12–S	12	6–S	8–S	12–S	5
5	2–(NR)	6–S	5–S	6–S	4–S	(NR)	(NR)	9–S	3
(NR)	7–M	5–M	6–M	9–M	6–V	2–M (NR)	6–M	(NR)	(NR)
(NR)	(NR)	(NR)	(NR)	(NR)	(NR)	(NR)	(NR)	(NR)	(NR)
(NR)	6–S	5–S	5–V	5–S	7–S	(NR)	4–V	8–S	(NR)
(NR)	1–V (NR)	5–S	5–S	3–S	2–S (NR)	(NR)	(NR)	5–S	(NR)
(NR)	1–V (NR)	5–S	5–S	3–S	2–S (NR)	(NR)	(NR)	5–S	(NR)
(NR)	(NR)	(NR)	(NR)	9–R	7–R	6	8	(NR)	(NR)
(NR)	(NR)	(NR)	(NR)	11–M	7–R	7	9	(NR)	(NR)
(NR)	(NR)	(NR)	(NR)	8	8	4	7	8–S	(NR)
(NR)	(NR)	(NR)	(NR)	3–S	2–S (NR)	(NR)	(NR)	5–S	(NR)
(NR)	(NR)	(NR)	(NR)	12–S	12	4–S	4	12	(NR)
(NR)	(NR)	(NR)	(NR)	7–M	6–M	(NR)	6–M	(NR)	(NR)
(NR)	(NR)	(NR)	(NR)	5–M	(NR)	(NR)	(NR)	(NR)	(NR)
(NR)	(NR)	(NR)	(NR)	5–S	6–V	(NR)	4–V	4–S	(NR)
(NR)	(NR)	(NR)	(NR)	4–S	3–S	(NR)	(NR)	5–S	(NR)
(NR)	(NR)	(NR)	(NR)	2–S (NR)	1–S (NR)	(NR)	(NR)	4–S	(NR)

A three-year life expectancy is the minimum requirement for recommendation.

Numbers indicate expected service in years.

A letter following a number indicates a primer must be used prior to application of compound or sealant.

choice on a price per performance basis. Each sealing job must therefore be handled individually.

An excellent summary of the properties of building sealants was conducted by J. Zakim and M. Shihadeh which was published in *Adhesives Age*.

5

Accessory Materials

Whereas the performance of a finished sealant depends largely on the properties of the base polymer, it depends also on accessory materials which are blended into the sealant. A sealant may contain adhesion promoters, fillers, pigments, plasticizers, thixotropic (antisag) agents, and other constituents. Various external accessory materials such as primers, release agents, and back-up materials are also used in order to effect a well sealed joint.

The elastomeric sealant is a sophisticated blend of many elements, but the average two-component material would contain the following ingredients as a minimum:

1. Base polymer
2. Filler, to control consistency and lower cost
3. Plasticizer, to modify hardness and modulus
4. Curing agent

5.1 Fillers

Fillers, in general, serve two purposes in a sealant: they are added to control the consistency or gunnability of sealants and, by adding bulk, they lower the percentage of base polymer required and thus lower the cost. However, the physical properties of the finished product can be markedly improved by the proper choice of fillers. The fillers most frequently used are carbon black, ground limestone (calcium carbonate), talc, clays, and ground silica. In nonsag sealants, for use in vertical joints, thixotropy is achieved by adding bentonite, asbestos fiber, or colloidal silica.

In general, the addition of fillers to a sealant increases the modulus of elasticity and lowers the ultimate elongation. Carbon black and ground silica are good reinforcing agents. Calcium carbonate and the clays are used to control consistency and have little effect on tensile strength or modulus. The aluminum fillers are pigments, rather than fillers, and produce a very soft sealant.

The particle size of the fillers is very important in the control of gunnability and thixotropy. Very fine particles are more easily dispersed throughout the mix and generally give the best reinforcement, especially in black systems.

Coal tar is frequently used as a filler in paving sealants because of its low cost, but it is of little value in building sealants.

5.2 Plasticizers

Plasticizers, when blended into a sealant, lower the modulus of elasticity and increase the ultimate elongation. Some plasticizers also improve the low temperature flexibility of the sealant. The plasticizer does what its name implies. It plasticizes the mix; that is, it makes it more workable. This increased workability also helps to release any air trapped in the sealant during the mixing process. Because of its effect on modulus and ultimate elongation, a plasticizer makes it possible to add high filler loadings to the sealant, which helps to reduce the cost.

5.3 Curing Systems

Sealants may be cured by chemical means, by catalytic action, by moisture absorption, or by solvent evaporation. In general, the two-component compounds are chemically cured, and the one-component compounds are cured by moisture absorption from the atmosphere or by solvent evaporation.

As the one-component sealants cure, they tend to "skin over" and cure from the surface inward. Once a tight skin has formed over the surface of the sealant, the cure proceeds much more slowly. For specification testing the one-component sealants are force cured at high temperatures and 100% relative humidity, whereas in the actual building joint a ½-× ½-in. bead of sealant may take several weeks to cure. The size of the sealant bead is therefore quite critical in the selection of a one-component sealant. The single component sealants may be a good choice for joints up to ⅜ in. wide. They are generally not suitable for wider joints, such as the ½-in. or larger joints which are often used with precast concrete panels. On the other hand, the two-component compound, since it cures more or less simultaneously throughout the mass, can be used quite successfully in wide joints as well as in the smaller ones.

The shelf life of all sealants is important. In the two-component materials, fillers may tend to settle out and cake after excessive storage time, which makes proper mixing very difficult. The one-component sealants, because of their cure system, are more seriously affected by long storage. Single component sealants furnished in the standard caulking gun cartridge may begin to cure, making them practically impossible to extrude. Although the useful shelf life varies for various sealants, it is probably safe to say that any material more than six months old should not be used in important work.

5.4   Solvents

It is advisable to keep the solvent content of most sealants as low as possible to avoid excessive shrinkage as the evaporation takes place. Silicones, polysulfides, and urethanes generally contain little or no solvent, while other sealants contain proportionately more. The two-component sealants are also more likely to have low solvent content. Single component sealants, especially the prepackaged ones, generally have a small amount of solvent added to improve gunnability. The principal solvents used are toluene, xylene, petroleum naphthas, and water. The consumer can spot the solvent content of a sealant quickly by reading the manufacturers' technical data sheets. A material indicated as "86% solids" will shrink in time to 86% of its initial volume. That is, the sealant contains 14% solvent or other evaporable materials.

5.5   Pigments

Sealants are offered in a wide range of colors to match any substrate. Some pigments will change the physical properties of the material as well as the color. The carbon blacks are both reinforcing and coloring agents. Titanium dioxide, in addition to being an excellent white pigment, gives moderate reinforcement. The aluminums yield a very soft sealant, and the best aluminum colors are formulated by adding a white pigment, such as titanium dioxide, along with the aluminum.

In general, however, any high quality pigment suitable for use in paint formulation may be used in compounding sealants. Approximately

1000 tons of pigments are used annually for this purpose. The silicones are offered in five standard colors, and the polysulfides in as many as twelve. Other colors can be provided on special orders.

5.6   Primers

The adhesion of a sealant to the substrate may be improved by either internal adhesion additives, primers, or by both. In general, the more elastic the sealant, the more it needs a primer to insure good adhesion. Priming of the joints is usually considered by the applicator as a necessary evil because it increases labor cost and constitutes one more step in the sealing process. However, Garden [9] states, "Priming . . . may be desirable in most instances if for no other reason than the fact that the joint surfaces are examined before installation of the sealant."

Since most sealant materials are relatively viscous, the purpose of the primer is to provide better wetting of the substrate. In essence then, the primer wets and, in a sense, seals off the substrate, and the sealant itself adheres to the thin layer of primer. Since the sealant actually adheres to the thin layer of primer, the primers used must necessarily be compatible with the sealant used. The primers most widely used are based on chlorinated rubber, silicones, urethanes, polyvinyls, and epoxy-polysulfides.

Although the primer must be compatible with the sealant, it must also adhere well to the particular substrate being considered. A typical urethane sealant, for example, might require one primer for use with metal joints and a different primer for use with concrete. In the case of joints of dissimilar materials, such as the joint between a concrete panel and a metal mullion, the better choice would be to use the concrete primer because good adhesion to concrete is much more difficult to obtain than adhesion to other substrates.

Primers may be either brushed or sprayed onto the substrate. Primers for porous substrates are best scrubbed into the pores of the joint wall with a stiff bristle brush, much in the same manner that portland cement paint is applied.

5.7   Back-up Materials

The primary purpose of the back-up material in the joint is to control the depth of sealant in the joint, and thus insure the proper shape factor. (See Fig. 3.1.) Another purpose of the backing is to provide support or reinforcement for the sealant material in horizontal joints, such as in floors and patios. Depending on the type of construction, the back-up material may be already in the joint: for example, the plastic or cork board joint former sometimes used in pavement construction.

The back-up material must be unaffected by any solvent contained in the sealant. Back-up materials containing asphalt or coal tar should never be used. These extrudable oils are incompatible with some sealants and may cause loss of adhesion. It is also quite possible that these extrudables may cause the staining of porous substrates.

Neoprene, urethane, and polyethylene foams, rubber or neoprene tubes, cork and fiber boards, and cotton rope and jute have all been used as backing materials. The foams have been the most successful materials because they are quite compressible with very little spread. The foams are readily available in strip form, in both round and rectangular cross sections to fit most joints.

It is relatively easy to place the back-up material at the proper depth in the joint. A simple depth gauge can be made by nailing a wood strip to the bottom of a cement finisher's wood float. This gauge can then be used to press the back-up material into the joint.

5.8 Release Agents

A release agent or bond-breaker strip should be placed into the joint on top of the back-up material to prevent adhesion of the sealant to the back-up. Strips of polyethylene film or tape make a very good bond-breaker. Special wax-backed tapes are available, but ordinary masking tape will do a passable job in an emergency. (See Figs. 3.1 and 3.7.)

5.9 Lubricant–Adhesives

The lubricant-adhesives are used only with preformed gasket seals. Depending on the type of installation, the lubricant-adhesive may serve only as a lubricant or it may serve both functions. In a typical glazing operation a simple soap solution may be used to ease the precompressed seal into the joint. However, in highway pavement joints, where preformed seals are widely used, both lubrication and adhesion are desirable. The lubrication function is necessary in order to place the seal in the joint properly; whereas the adhesion function is desirable in order to help the seal maintain its proper position in the joint as the pavement moves.

Considering the two functions, then, it becomes apparent that different materials are used as lubricant-adhesives. In applications where only lubrication is required, a soap solution or other nonoily film may be used. Where some degree of adhesion is desired, a thin layer of any compatible sealant, such as a neoprene, may be used.

6

Installation

As might be expected, the various types of sealants require different installation techniques. Since the installation technique is a function of the form and manner in which the sealant is supplied, we list the types of sealants which might be furnished to the job site.

Single component mastics
Two-component mastics
Tapes
Preformed gaskets
Foams

6.1 Mastic Sealants

The one- and two-component mastics differ in both storage and preparation of the sealant for use, but the basic installation procedure is the same for both types.

First the joint is cleaned, the back-up material is placed in the joint, and the bond-breaker strip is placed on top of the back-up. In order to insure a neat joint, masking tape may be placed on the substrate along both exterior edges of the joint. The joint is then primed; and when the primer is just dry to the touch, the sealant is extruded into the joint from hand- or air-operated caulking guns. Figure 6.1 shows a sealant being installed with an air gun. The surface of the joint is then tooled to force the sealant into the joint and insure better wetting of the substrate, and to provide the proper contour to the sealant surface. Quite often, slick-

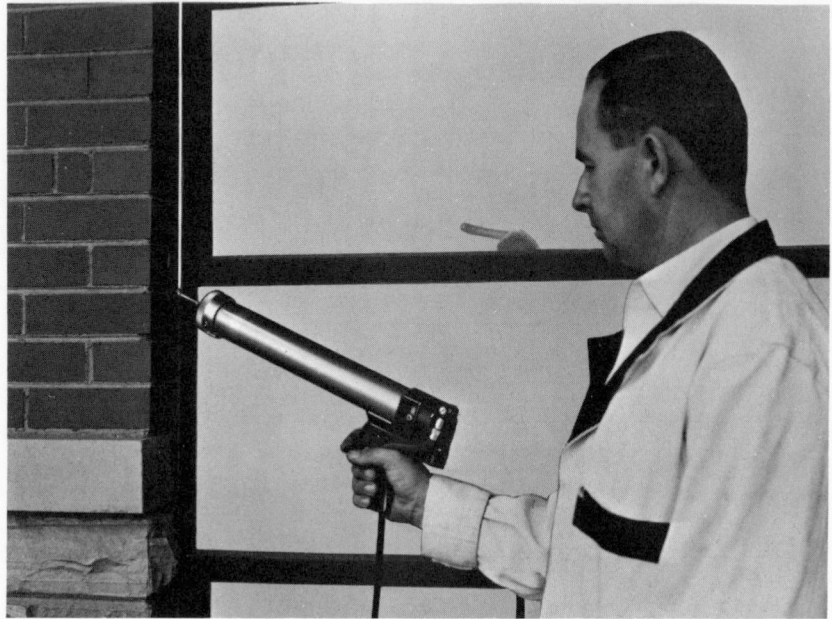

Fig. 6.1 Sealant application using an air-operated caulking gun. (Courtesy of Pyles Industries)

ing agents such as solvents or soapy water are used to make tooling easier. After the tooling is completed, the masking tape along the sides of the joint may be removed.

In order to provide the best installation, the caulking gun should be pushed, rather than pulled, along the joint. This forces the bead of sealant out ahead of the nozzle and does a better job of filling the joint opening. Pulling the caulking gun along the joint without tooling does give a neater appearing joint, but is more likely to leave voids in the body of the sealant.

Weather is also an important factor in joint seal installation. Many of the two-component mastics will not cure at temperatures below 45 F. The higher the ambient temperature, the faster the curing rate. The curing of the single component mastics also depends on temperature and humidity. A warm, humid climate accelerates the cure.

If one-stage weatherproofing is being employed in a concrete building, rain can slow down a sealing operation for several days because a satisfactory method has not yet been found for priming or sealing saturated concrete (15 to 18% moisture content). In some cases, standing water accumulates in horizontal joints thereby making sealing impossible.

Fig. 6.2 Air-operated caulking gun adaptable for either cartridge or bulk loading. (Courtesy of Pyles Industries)

6.1.1   Single Component Mastics

The single component mastics are furnished to the consumer in $\frac{1}{10}$-gal cartridges, or in 1- and 5- gal pails. The cartridges are clean and easy to handle, store, and use. The cartridges are simply inserted into a manual- or air-operated caulking gun, the plastic tip is then snipped off, and the unit is ready for use.

Bulk material in the larger pails is applied with a bulk-loading caulking gun. The bulk-loading gun may be either a suction type, which is loaded by dipping the nozzle into the sealant and withdrawing a plunger, or a rear load type, which is filled by spatula from the larger pail.

Figure 6.2 shows an air-operated caulking gun which is adaptable for either cartridge or bulk-loading use. Figure 6.3 shows special equipment for power-operated bulk loading. Figure 6.4 is a portable air unit which the caulking gun operator can conveniently carry onto a scaffold or to isolated job locations. The air unit is strapped to the operator's body, leaving both hands free for work.

The advantages of the single component sealants are that they require no on-site mixing and maintain a longer working life. This longer working life is of value especially when the contractor is using material furnished to the job in the larger containers. The contractor may open a 5-

Fig. 6.3 Power-operated bulk loading. (Courtesy of Pyles Industries)

gal pail of sealant and refill guns from it over a period of several hours. It is also possible to open a bulk container one day and still use the material the following day. There may be some skinning over of the material in the opened pail, but this skin may be removed and discarded, and work can proceed.

6.1.2   Two-Component Mastics

The two-component sealants are installed into the joint by using the same types of caulking equipment as for the single component sealants. An additional requirement of the two-component materials is that they be thoroughly mixed at the job site. The two-component materials may be furnished in the same container sizes used for the single component sealants, but the largest percentage of material is supplied in 5-gal pails. These materials are always supplied to the job site in premeasured quantities, so that if all the material in container A is mixed with all the material in container B, satisfactory proportioning of components is assured. The two components of the sealant are generally furnished in different colors so that the operator can tell when the mixing is complete. The great bulk of two-component sealants are mixed at the site, using a paddle on a slow speed electric drill (450 rpm). Faster mixing, which whips excess air into the mix and also heats up the sealant, would shorten the working life.

Excellent construction kit mixers are available which use a rotating screw to mix the material. These mixers, although expensive, do a good job of mixing without entrapping air in the sealant. After mixing, the apparatus (Fig. 6.5) ejects the material so that it can be conveniently loaded into bulk-type caulking guns.

The two-component sealants which are used in highway and airfield pavement joints may be installed with special equipment. The two components of the sealant are furnished in a 1:1 volume ratio. They are pumped separately through hoses, intimately mixed at the nozzle, and extruded into the joint. Then, this entire unit is truck or trailer

Fig. 6.4 Air gun with portable air supply. (Courtesy of Semco Sales and Service)

Fig. 6.5 Mixer for two-component sealants. (Courtesy of Semco Sales and Service)

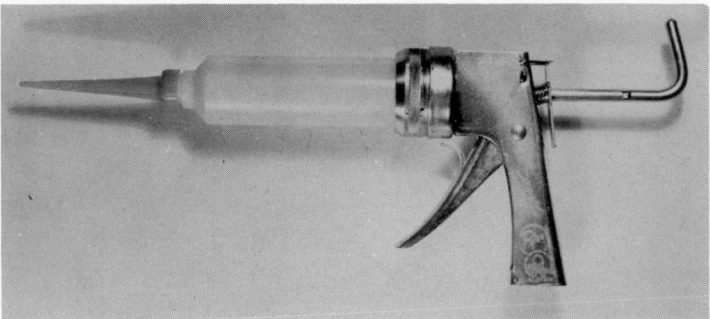

Fig. 6.6 Hand gun for use with premixed and frozen sealants. (Courtesy of Pyles Industries)

mounted in order to cover the distance involved in pavement sealing.

The cartridge-supplied two-component sealants are specialty items and have received little acceptance by contractors. There are two systems by which these materials may be formulated. One system furnishes a kit with two plastic tubes: one containing the base material, and a smaller one containing the curing agent. The curing agent is injected into the other component, and the material is then mixed in the tube. This system has the obvious difficulty of there being no way for the operator to know when the material is thoroughly mixed.

The alternate system for furnishing two-component materials in cartridges is to premix the components at the factory, load the sealant into polyethylene cartridges, and freeze the sealant to retard the curing process. The cartridges are then thawed at the job site and used. The six-month shelf life limitation is especially important with this type of packaging. Figures 6.6 and 6.7 show the hand and air guns often used with the frozen sealants.

The two-component materials which are site mixed have a definite pot life limitation (usually 2 to 3 hr). The contractor should mix only as much material as he can use within this time period. Any excess material cures into a block of rubber, and cannot be reworked.

6.2 Tapes

The sealant tapes, to put it very simply, are mastic-type sealants which have been thickened enough to be furnished in roll form. They are generally supplied with a paper backing which is removed immediately prior to installation.

Being modified mastic sealants, they are subject to some of the same

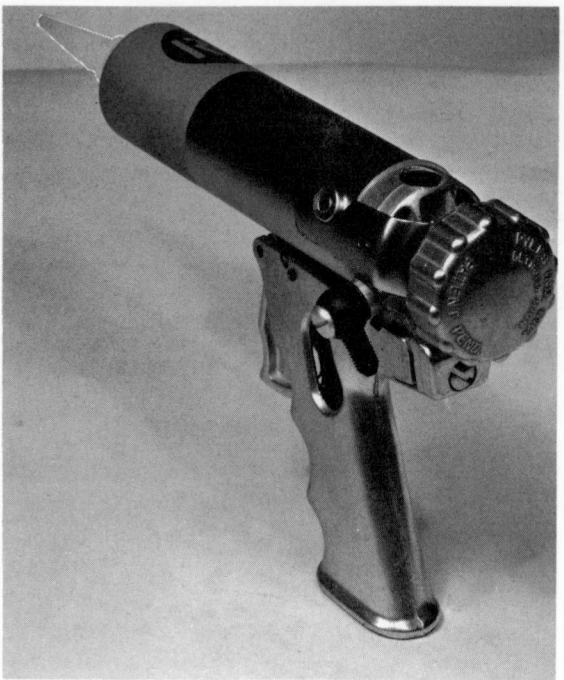

Fig. 6.7 Air gun for use with premixed and frozen sealants. (Courtesy of Pyles Industries)

installation procedures as the gun-applied mastics. The joint opening should be cleaned. The tapes are formulated from lower recovery polymers and thus joint cleaning is not so critical as with the higher recovery materials; but joint cleaning should not be omitted. Since the tapes are usually low recovery sealants, priming of the joint is seldom necessary. The actual installation of the tape sealant is accomplished by hand. Strips of tape of the proper length are cut from the roll with an ordinary pocketknife or scissors, and the sealant is pressed into place. Butt splices and corner splices are easily formed by pushing the tape into place with the fingers. Because of the types of applications, joint tooling or finishing is seldom necessary.

6.3   Preformed Gaskets

The preformed gasket seals are used in literally hundreds of specialized applications and, consequently, are supplied to the job site in a multitude of shapes. Common to all these shapes is the fact that they are installed under compression. Since adhesion is no problem, joint cleaning may consist of simply blowing or brushing loose dirt out of the joint. A lubricant is wiped along the gasket, and then the gasket is pressed into the joint. In exterior joints, such as with precast concrete or metal curtain wall panels, the building panels are erected first and then the seal is compressed and placed in the joint. In a glazing application the installation will vary according to the type of window unit, but the following sequence is typical.

1. Place the gasket on the window framing.
2. Set the glass either next to or into the gasket, depending on type.
3. Fasten the inside stop strip.

The critical factors to be considered in gasket installation are stretching and splicing. The gasket must be designed for the joint. In general, compressing the seal about 40% during installation is desirable. If the gasket is stretched longitudinally during installation, it decreases in cross section and does not exert sufficient interface pressure to function properly. The gasket sections must also be very carefully cut to length for installation because there is, at present, no adhesive which will form a satisfactory structural splice.

No joint finishing is required with these dense extrusions because they furnish an excellent finished appearance.

6.4 Foam Sealants

The closed-cell foam seals have at least as many possible applications as the dense gasket extrusions and, consequently, are available in a wide variety of shapes. The foam seals are also a compression-type seal, and the installation process is much the same as for the dense extrusions. Required cleaning of the joint is minimal, since the spongy nature of this type of seal permits it to flow around and into irregularities in the joint face. Priming of the joint is not required. The seal should be installed under a compression of approximately 25%. Compression beyond 50% is not recommended. Compression above this level distorts the foam structure excessively and causes a drop in the pressure which the seal exerts against the joint wall. The foam seals should not be stretched longitudinally during installation because this reduces cross section and, consequently, reduces sealing pressure. Butt splices can be made satisfactorily in the field using a suitable adhesive. Furthermore, no tooling or finishing of the joint is required when foam seals are used.

6.5   Joint Cleaning

Proper cleaning of the joints is undoubtedly the most tedious step in the entire sealing process, but its importance cannot be overstated: no mastic sealant will adhere to a dirty joint wall. Joint cleaning is somewhat less critical with the preformed shapes, but it is still required. Loose paint, scale or rust on metal surfaces, laitance on the face of concrete joints, and general construction dirt must all be removed before the joint can be sealed. Sound metal surfaces, such as new aluminum, can be cleaned with a nonoily solvent and a soft rag. Scale, rust, and laitance are best removed by sandblasting. A clean wire brush can also be used for this purpose. In resealing work, old sealant can be cut out with a knife, and the remaining traces of old sealant can be removed from the joint faces by sandblasting or wire brushing. After sandblasting or wire brushing, the joint faces should be dusted clean with a soft brush or rag, or by vacuum. Compressed air should not be used to blow out the joints because oil from the air compressor is often sprayed onto the surface. The degree of cleanliness required by the mastic sealants is almost in direct proportion to the amount of recovery in the sealants. The silicones and urethanes require very clean joints; whereas, at the other end of the spectrum, the PVA ° latex caulks and solvent-based acrylics require much less.

Aluminum mullions and sash, as furnished to the job site, present a special cleaning problem. The aluminum sections are often given a special lacquer finish at the factory. This lacquer coating should be removed so that the sealant can adhere directly to the aluminum. A strong solvent, such as xylene, on a clean rag can remove the lacquer.

6.6   Heaters

The solvent-based acrylic sealants are generally thermoplastic, and require heating to 120 F before they can be extruded satisfactorily from a caulking gun. This heating can be accomplished in a special heating chest available from the sealant manufacturer. Some contractors have built their own heating units: often nothing more complicated than a wooden box with a light bulb for a heating element.

The same type of chest is sometimes used to thaw premixed and frozen cartridges of two-component sealants. Heaters must be used with

° Polyvinyl Acetate.

great care in this application, because heat accelerates the cure of the two-component sealants. If too much heat is applied, some of the material may begin to cure before the rest of the sealant in the cartridge has had time to thaw out to a gunnable consistency.

In the machine application of two-component highway pavement sealants, special heaters are sometimes used to control the viscosity of the sealant components so that they can be easily pumped through the hoses to the nozzle.

In some highway and airfield applications, hot poured rubber-asphalt sealants are used. These materials require special double-wall melting kettles to heat the sealant to about 375 F.

7

Polysulfide Sealants

7.1 Introduction

The polysulfide sealants were the first elastomeric joint sealants to be used in building construction. Because these sealants have had the longest record of experience and since they still occupy the largest single segment of the elastomeric sealant market, they are quite often the standard by which other sealants are judged.

The polysulfide rubber was first produced by Thiokol Chemical Company in 1929. As with many new products, the discovery of the polysulfide occurred as an accidental by-product of other research. In the course of another study in 1929, the investigators noted a rubbery mass accumulating as a waste by-product. Recognizing the possibilities, the researchers refined and improved this waste mass into a usable, synthetic rubber.

This new synthetic rubber was developed slowly until the World War II years when the shortage of natural rubber forced the development of the synthetics. Vigorous growth during the war years and afterward led to a multitude of uses for the new rubbers; and in 1952 the polysulfide sealant was introduced to the construction industry.

Currently there are about 35 formulators of two-component polysulfide sealants; ten of these are of major size. Thiokol Chemical Company, which has long held the basic patents to the liquid polysulfide polymer, is virtually the sole supplier of the polymer to all formulators, although other suppliers are now entering the market.

The polysulfide sealant is the only elastomeric sealant which has a nationally accepted specification containing its properties (USASI 116.1,

SS-S-00200). Specification sealants for important work generally contain approximately 40 to 50% polymer. Sealants for nonworking or minimal movement joints may contain higher filler loadings and lower polymer content. Because of the number of formulators involved and the intense price competition, the polymer content and, consequently, the quality of the polysulfides began to drop considerably in the early 1960's. Recognizing the problem quickly, Thiokol instituted a hallmark program, under which they test the finished products of their formulators and issue a seal of approval. This program has been so markedly successful as a quality control measure that other raw material suppliers are giving it serious consideration.

7.2 *Compounding*

The polysulfide polymer is a thick liquid with the color and consistency of honey. This polymer can be cured into a rubber by adding $7\frac{1}{2}$ parts of lead dioxide paste to 100 parts of polymer. This ratio is too unbalanced for easy mixing, hence the curing agent is generally bulked-up to 15 parts per hundred. The rubber formed from this pure polymer has very poor properties and especially poor tear strength, so the mixture is strengthened by the addition of reinforcing fillers. A plasticizer is usually added to make the mixture more workable and lower the modulus of elasticity. The finished sealant would probably also contain a minor percentage of an adhesion additive and about 1% of a mild acid to retard the set and improve the pot life. The composition of the major components of the sealant might be as shown in Table 7.1.

Table 7.1

LP (liquid polymer)	100 parts
Reinforcing filler	50 parts
Plasticizer	50 parts
Curing agent	15 parts

The curing agent component could be increased in bulk from $7\frac{1}{2}$ to 15 parts by incorporating some of the plasticizer into the lead dioxide paste. The finished sealant would then arrive at the job site in two components: component A containing the polymer, filler, and most of the plasticizer, and component B containing the lead dioxide paste and the remainder of the plasticizer.

Fillers, in general, strengthen the polymer up to a point. Too much filler loading, of course, simply results in excessive dilution of the polymer and hence a poor quality sealant. Any filler loadings which reduce the polymer content to below 25% may be considered excessive. The carbon blacks, titanium dioxide, calcium carbonate, ground silica, and hydrated alumina are used, among others, as fillers for polysulfides. As mentioned before, acids tend to retard the cure; therefore the clay fillers, which are usually acidic, are not the most desirable. Coal tar is used as a filler for highway sealants, but its only benefit is to lower the cost. Because of intense competition, compounding data are jealously guarded and thus the consumer has no way of knowing the ingredients of any particular sealant. The consumer must then rely on published test data to evaluate the material.

As with most elastomeric sealants, fillers increase the modulus of elasticity and lower the ultimate elongation. Plasticizers have the opposite effect: they decrease the modulus and increase the elongation. Consequently, with the proper balance of filler and plasticizer it is possible to custom-blend the finished sealant to furnish a desired set of physical properties.

The polysulfides can be compounded in a wide range of colors. The base polymer is an amber liquid, and the lead dioxide is a brown paste; therefore pure cured polysulfide has a light brown color. Most pigments work quite well with polysulfides, although not exhibiting the versatility that is available with the clear and colorless silicones. If necessary, the polysulfides can be cured with other curing agents, such as manganese dioxide or some organic peroxide, consequently colorability is no problem.

Strictly speaking, the two-component polysulfides are catalytically cured; but they are generally considered a chemically cured sealant. The two components, as furnished, will cure to a rubber at room temperature after being mixed. The sealant must be well mixed in order to function properly, but there is a slight built-in safety factor in the polysulfides because at least some amount of diffusion cure will act in the user's favor in the event of incomplete mixing.

The one-component polysulfides were first introduced to the construction market in about 1961. The single component polysulfides cure by absorbing water from the atmosphere. As with the silicones, this reaction results in a tough rubbery skin forming at the surface which slows down the cure rate for the remainder of the sealant mass. Since polysulfides are normally used in wide joints requiring large beads of sealant, the one-component materials were at first used in the same type of application. Early results with the single component material were not as

Fig. 7.1 Social Science Building, University of California campus in Los Angeles, weatherproofed with a one-component polysulfide sealant. (Courtesy of Products Research and Chemical Company)

successful as with the two-component material and thus acceptance was slow at first for the one-component sealant.

Manufacture of the single component sealant is more difficult and expensive than its two-component counterpart. Not all formulators are equipped to compound the single component material. However, a well compounded single component sealant will meet the Federal specification, and only recently Thiokol has issued its first seal of approval to a single component polysulfide. The Social Science Building at the University of California, shown in Fig. 7.1, is an example of a structure weatherproofed with this material.

The polysulfides are available in either a gun grade for vertical joints or a self-leveling grade for horizontal joints, as in pavements. The gun grade material contains a thixotropic agent so that it will not sag or flow out of wall joints. Bentonite or asbestos fiber could be used. The self-leveling sealants may have a small amount of solvent added to improve the flow, but, in general, the polysulfides are virtually 100% solids and thus there is no shrinkage after the sealant has been placed in the joint.

The polysulfides can successfully be blended with other polymeric materials to produce specialized sealants and coatings for construction use. Polysulfides serve particularly well as a flexibilizer for the epoxy resins. Polysulfide epoxy materials, which utilize the high strength and abrasion resistance of the epoxies and the flexibility of the polysulfides, make an excellent protective coating for concrete bridge decks and parking garage ramps. Epoxy polysulfide sealants have been formulated for specialized use, but have received little attention in the overall sealants market.

7.3   Properties

It is difficult to state quantitative values for a given set of properties associated with the polysulfides, because they have been compounded for a wide variety of uses. Also, with about 35 formulators competing to produce a high quality, salable sealant, it is natural to expect variations in the properties of the finished product. The values given here may be considered as representative of high quality building sealants which meet the requirements of the Federal specification.

7.3.1   Odor

The polysulfides in the uncured state are noted for a very disagreeable odor; therefore adequate ventilation is recommended when working with these sealants. However, this odor disappears as the sealant cures, hence the finished product in the joint is virtually odor-free.

7.3.2   Toxicity

Extensive laboratory tests have shown that the polysulfide sealants are nontoxic and nonallergenic. Nevertheless, it is advisable to avoid prolonged contact with the skin. The polysulfides do, however, require the use of a solvent to clean the operator's hands after use, and repeated solvent washing can cause cracking of the skin.

7.3.3   Solvent Resistance

The polysulfides are one of the better performers with regard to solvent resistance. They have good water immersion resistance; this makes them prime candidates for exterior, one-stage sealing. One of the Federal specifications covering airport pavements contains a jet fuel resistance requirement, and the polysulfides regularly meet this specification. These sealants also have very good resistance to other organic solvents, oils, and a wide range of chemicals.

7.3.4 Hardness

The hardness of polysulfide sealants varies from virtually zero up to values in excess of 50 on the Shore scale. This versatility is partially due to the amount and type of filler loading. Another factor affecting hardness is the base polymer itself. The polysulfide liquid polymer is not merely one, but rather a whole family of polymers. The amount of cross-linking in the polymer determines a set of properties. Materials with very little cross-linking between the polymer molecules are very soft. As the cross-link density increases, hardness also increases. A typical coal tar extended highway sealant has a hardness of about 5, but typical building sealants have a hardness in the range of 10 to 30. For terrace and patio joints, a relatively stiff (25 to 30) sealant should be used to prevent puncture of the sealant by stones and spike heels. In other building joints, the entire range of hardness values is open, except that sealants in low, exposed locations should be toward the high end of the range to prevent picking and gouging of the sealant by vandals.

Hardness of the polysulfides does vary with temperature. A typical building sealant might vary from 10 to 120 F to a value of 40 at −40 F. It is worth noting, however, that the hardness values of polysulfide sealants tend to converge at cold temperatures. Thus two sealants with respective Shore values of 10 and 25 at room temperature would share approximately the same value of Shore hardness at very cold temperatures.

7.3.5 Aging and Weathering

The polysulfide sealants have good resistance to aging and weathering. They do not have as good weathering characteristics as the silicones, but the polysulfide sealants will successfully stand up after an exposure of more than 1000 hr in the weatherometer; they may therefore be considered very good for exterior use. Many of the earlier polysulfides which were used in highway pavements showed some surface crazing or "alligatoring" after exposure to road dirt, traffic, and highway salt. Most of the polysulfide sealants currently formulated for highway work have shown a great improvement in surface appearance.

7.3.6   Ultraviolet Resistance

The polysulfide sealants have a UV resistance sufficient for use in glazing work. The weatherometer test includes UV exposure as part of its cycle, and the polysulfide sealants meet this test. Also, there are enough glass curtain wall buildings performing well to attest to this property.

7.3.7   Modulus of Elasticity

The modulus of elasticity of polysulfide sealants is not a constant quantity. Since the polysulfides are a relaxing type of sealant, the modulus depends on many factors. The modulus can be increased by using a polymer of relatively high cross-link or by increasing filler loading; and it can be decreased by adding plasticizer. Specimen shape and testing rate also affect the value of modulus. One of the Federal specifications calls for extending the sealant specimen at a rate of ⅛ in./hr. This testing rate is practical for concrete bond specimens, but virtually useless for determining a modulus of elasticity for the polysulfides. The polysulfide sealants have a stress relaxation curve which is steeper than exponential for very short time values. Consequently, the stress relaxes to virtually zero by the time the specimen has been extended ⅛ in. At the other end of the time scale, the British Building Research Station has shown that aluminum substrates move at 120 in./hr or 2 in./min. This testing rate can be duplicated in the laboratory, but only with some difficulty. Fortunately, the polysulfide sealants have excellent adhesion to aluminum, even without a primer. Between these two extremes of testing rates lies the great bulk of sealant applications, and a testing rate of ⅛ in./min. is considered as representative of most "slip and stick" building movements. With this testing rate, most polysulfide building sealants show a modulus of elasticity of 30 to 80 psi at 50% strain. This low value of modulus indicates that the polysulfide sealant is sufficiently compliant to be used in concrete joints without causing any spalling failures of the concrete substrate.

7.3.8   Ultimate Elongation

Polysulfide sealants can be formulated for a wide range of ultimate elongation values by varying the polymer, the filler loading, and the plasticizer. Specimen shape also affects elongation, but even in the ½-×½-×6-in. specimen, which is representative of field conditions, wide variations are possible. Laboratory specimens of polysulfide with elongation values of over 1000% have been compounded, but these materials have virtually no shape recovery from such excessive deformations. Practical sealants for use in building joints usually have ultimate elongation values in the neighborhood of 100%.

7.3.9 Creep and Stress Relaxation

These two properties define the amount of internal flow which takes place in the sealant. The polysulfides occupy an intermediate place on the scale: between the high recovery silicones on one hand and the solvent-based acrylics on the other. A polysulfide sealant when extended 50% will flow internally and relieve the stress by as much as ⅓ in the first 20 min. This property can be both an advantage and a disadvantage. The stress relief, of course, is the advantage, but it is necessarily accompanied by a corresponding lack of recovery. If subsequent joint movements are short increments, the result is a wrinkling of the sealant surface. If subsequent joint movements are large, the result may be a failure of the type shown in Fig. 3.15.

Table 7.2 lists quantitative values for the physical properties of a typical polysulfide building sealant.

Tables 7.3 and 7.4 are summaries of the properties of the one- and two-component polysulfides as observed by the architects who specify the sealants and the contractors who use them.

Table 7.2

Tensile strength	180 psi
Ultimate elongation (½- × ½-in. specimen)	100 %
Tear strength	80
Hardness, Shore A	20
Adhesive strength	180 psi

7.4 Application

The two-component polysulfide arrives at the job site in a preproportioned kit. The curing agent is mixed with the base material immediately prior to use. A pot life of 3 hrs is normal, but this can be adjusted at the factory to as short as 10 min for horizontal joints which must carry foot and wheel traffic. For building joints the material is usually hand mixed by using a paddle and slow speed drill. This mixing is messy and expensive, and is the reason that many architects prefer the one-component materials. The two components of the sealant are fur-

Table 7.3 Advantages and Disadvantages of One-Component Polysulfide Sealants
(as reported by cnnsumers)

Advantages	Disadvantages
1. One-component system	1. Require a moisture cure (low humidity, slow cure)
2. Broad range of cured hardnesses available	2. Require primers
3. Good durability	3. Inferior elongation and flow properties more difficult to control than the two-component polysulfides
4. Withstand large joint movement without failure	4. Recovery (high compression) not as good as silicones and not available in as many colors
5. Low shrinkage	5. Very slow skinning and cure rate (10 to 20 days for complete cure), therefore they remain tacky and pick up dirt, and children can pull them out of the joint prior to complete cure
6. Good adhesion	6. Cannot be applied at temperatures below 40 F
7. Meet Federal specification TT-S-00230	7. Complete exclusion of moisture during manufacture is essential
	8. Poor package stability (shelf-life)
	9. Obnoxious odor

nished in different colors so that the operator knows when the material is sufficiently mixed. The John Hancock center, shown in Fig. 7.2, is an example of a structure in which the two components were premixed and frozen.

The polysulfide sealants extrude well from any standard caulking gun. They are sufficiently flowable to fill any voids or irregularities in the substrate. Joints can be tooled with a pointing tool or small spatula dipped occasionally in solvent.

The polysulfide sealants neither mix well nor cure well at low temperatures. They should not be placed at temperatures below 40 F.

Polysulfide sealants require clean joints in order to function properly. Joint cleaning is not as critical as with the silicones, but the joint cleaning cannot be omitted or done haphazardly.

Polysulfide sealants will adhere well to steel, aluminum, glass, and porcelainized panels without the use of a primer. A primer is required for wood and concrete. However, to insure the best performance,

Table 7.4   Advantages and Disadvantages of Two-Component Poly-sulfide Sealants
(as reported by consumers)

Advantages	Disadvantages
1. Especially good adhesion for metal-to-metal and concrete joints	1. Require messy mixing; if not correctly mixed will not perform properly
2. Superior elongation compared to the one-component polysulfides	2. Cost $2–$4 per gallon labor charge to mix and place into the caulking gun
3. Good water immersion resistance	3. Poor cure over moist surfaces (moisture destroys bond prior to complete cure)
4. Do not require as critical surface preparation as silicones	
5. Good flexibility, elasticity, and elongation	4. Poor low temperature cure (not below about 40 F)
6. Fast cure so they can be painted over and walked on (spiked heels) in a short period of time	5. Inferior recovery (high compression set) relative to polyurethanes
7. Well balanced cohesion and adhesion properties	6. Colorability not as good as silicones
8. Good durability or long life (20 to 30 yrs) in moving joints	7. Penetration resistance not as good as polyurethanes in traffic-bearing areas
9. Survive large movements (as much as 50% of the joint width) without failure	8. Require more surface preparation than solvent-based acrylics
10. Good resistance to ultraviolet light	9. Pot life and cure rate dependent on temperature and humidity (high temperature and humidity accelerate cure and vice versa)
11. Resistance to heat, chemicals, oil, and solvents	
12. Nonstaining to masonry	
13. Negligible shrinkage	10. Unpleasant odor
14. Soft and flowable (before cure) to completely fill irregular openings	11. Primers required with masonry (concrete, stone, marble) and wood
15. Cure chemically without contact with air	
16. Broad range of cured hardnesses are available	
17. Primers not required with glass, porcelain, ceramics, and metals	
18. Meet Federal specification TT-S-227b	

Thiokol recommends the use of a primer for any sealant which is to be used in working joints.

The polysulfide sealants are competitively priced among the premium sealants, but are still well above the price of oil-based caulks and butyls. Consequently, they are most often used in working joints or exposed lo-

Fig. 7.2 The John Hancock Center. Much of the sealant on this job was a premixed and frozen two-component polysulfide. (Courtesy of Products Research and Chemical Company)

cations where their properties can be utilized (Fig. 7.3). They are not often used as bedding materials or in nonworking joints.

7.5 Summary

The polysulfides are competitively priced, high quality sealants. They have a good balance of adhesive and cohesive strength, and good tear resistance. They are available in a wide range of colors and hardness values, and can tolerate joint movements of up to 50%. They are available as one- or two-component materials and in both a flowing and a nonsag grade. They have good resistance to aging and weathering over a relatively long life span. Their recovery is not nearly as good as either the silicones or the urethanes. They should be used with a primer in all working joints, and they must be applied into clean joints in order to function properly.

At present the polysulfides are well established in the construction industry and, excluding the oil-based caulks, they account for slightly less than half of the construction sealant market. However, the increasing ac-

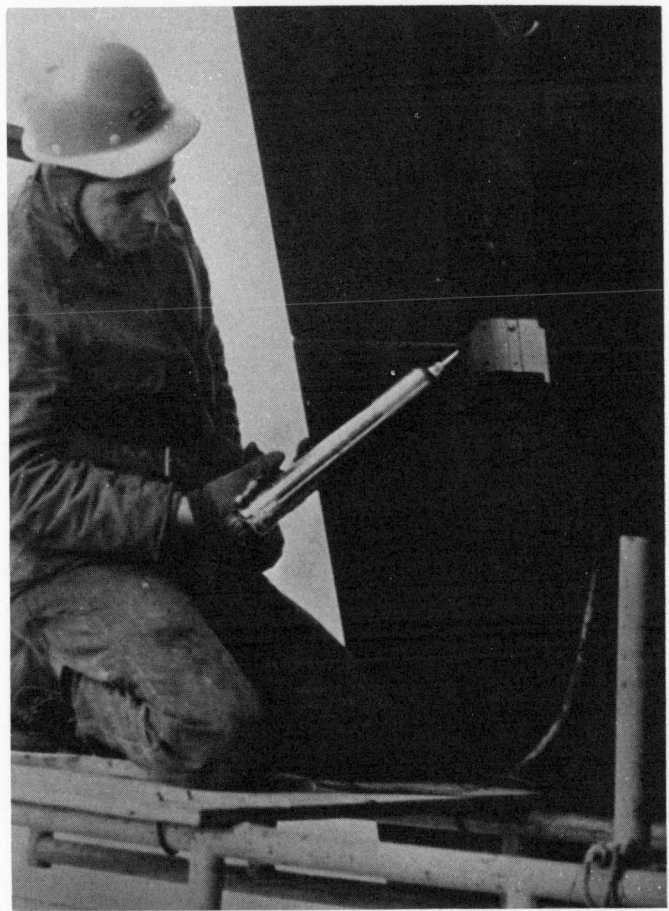

Fig. 7.3 The John Hancock Center. Sealing exterior bracing joints with a factory-mixed, prefrozen polysulfide sealant. (Courtesy of Products Research and Chemical Company)

ceptance of silicones and urethanes, together with the growth of two-stage weatherproofing, should greatly alter the picture. The polysulfide sales will continue to grow, but they will probably not maintain their present high percentage of the market.

8

Silicone Sealants

8.1 Introduction

Although the roots of silicone technology date back to the mid-1860's, it took nearly a century for the first silicone sealants to become available to the construction industry. Much of the early research work on silicone compounds was directed toward the development of electrical insulating materials. In the early 1940's the first silicone elastomer was developed, and this rubber found some applications in the aircraft industry. In 1960 the silicone sealants were introduced to the construction industry.

The silicone polymer is quite versatile and is being used in hundreds of applications including sealants, adhesives, potting compounds, coatings, and molded extruded parts. The sealants consumption is, as yet, a small but growing portion of the total silicone market.

At the present time there are two major manufacturers of silicone construction sealants: General Electric Company and Dow Corning Corporation. Two other major manufacturers of silicone products, Union Carbide Corporation and Stauffer Chemical Company, have shown interest in construction sealants, but have not as yet made any noteworthy penetration of the sealant market.

It is significant that both of the major suppliers of this material market a finished sealant product. They do not furnish the basic polymer to formulators for compounding. Consequently, these producers are able to exercise in-plant quality control and thus the result is an end product which is quite uniform from batch to batch and throughout the construction industry.

The silicone sealants are by far the highest priced construction sealant. This high price, together with the need for priming and critical
joint preparation, has held the silicones to a steady but modest growth
in the sealant market.

8.2 Compounding

The basic silicone polymer, clear and colorless, is obtained by the reduction of silica sand. Many of the same fillers that are used for other
elastomers are also used in silicone sealants: these include calcium carbonate, clay, and ground silica. These fillers tend to increase modulus
and lower extensibility with silicones as well as with other sealants.

Since the basic polymer is clear, the range of color tints is unlimited.
Both manufacturers market the sealant in five standard colors, but special colors may be supplied in large orders.

The finished sealant in the uncured state is quite soft and workable
without the addition of solvents. Consequently, the one-component sealant is almost 100% solids, which means negligible shrinkage after the
sealant is placed in the joint. This absence of solvent also helps to give
the silicone sealants a shelf life somewhat better than that of many other
one-component materials.

8.3 Properties

The outstanding property of the silicone sealants is high recovery. Specimens compressed and held for one year may show as much as 98% recovery upon removal of the load. Since most of the other properties of
this material can be duplicated by other sealants, the selection of a silicone sealant by a prospective user often hinges on whether the user is
willing to pay the high price in order to obtain this high recovery.

In the uncured state, the silicone sealants have an excellent gunning
consistency and very good shelf stability, which make them popular
with the craftsman on the job. An opened cartridge of one-component
sealant need not be used all at once. The operator has to force only a
little bit of the sealant into the plastic tip of the cartridge and then let it
cure. This plug of cured sealant effectively closes the opening. To reuse
the cartridge, this small plug of cured material can be removed and discarded, and work can proceed.

Another factor which makes the silicones popular on the job is their
exceptionally stable viscosity. Temperature has little or no effect on the
gunning characteristics of the sealant. The silicone sealants can be easily

extruded from the caulking gun over a temperature range of −40 to over 200 F. Cure, of course, is retarded by cold weather, and therefore low temperature installation is not generally recommended.

The curing of the silicone involves acetic acid; consequently, the uncured material gives off a strong vinegar smell. Since this can be irritating, adequate ventilation is recommended when working with silicones.

For larger joints, the silicone sealants are available in a two-component sealant, but this material has not received nearly as much acceptance as its one-component counterpart.

In addition, General Electric has quite recently produced a lower recovery sealant. This material has just been introduced, and information on its performance is not yet available.

In the cured state the silicone sealants are a very stable high recovery sealant. Shore hardness will seldom vary more than five points over a temperature range of −40 to +180 F, even after extended exposure. A typical silicone sealant will have a Shore hardness of about 35.

These sealants, however, are characterized by low tear resistance, which is partly due to the high recovery. A typical silicone sealant has a tear resistance of 40 lb/in. in comparison with 70 to 80 lb/in. for the polysulfides and polymercaptans. In practical terms, this means that it takes about 40 lb of force to propagate a tear in a notched and stressed specimen. Because the silicone is a high recovery sealant, it remains under stress when the joint is deformed. In service this means that once the deformed sealant is cut or punctured, the existing internal stresses will extend the tear very rapidly.

The silicone sealants exhibit excellent resistance to weathering. They are not affected by ozone, which is a reactive form of oxygen. The silicones have excellent resistance to moisture and ultraviolet radiation. One thousand hours in a weatherometer is considered a practical minimum for sealants exposed to the elements, however a typical silicone sealant can be exposed several times this minimum without any significant change in its properties. This excellent weathering resistance makes the silicones prime candidates for exterior one-stage sealing.

The silicones sealants in the cured state are somewhat stiff. The Shore hardness of 35, although stable, is quite close to the practical upper limit of 40 for building sealants. This stiffness is also reflected in the tensile strength value of 400 psi, which is a good average value for silicones. In the $\frac{3}{8}$-×$\frac{3}{8}$-in. joint considered typical for exterior sealing applications, this high tensile strength would mean that a force of 1800 lb per linear foot of joint would be exerted on the adhesive interface while the joint remained extended.

The tensile adhesive strength is somewhat low (150 psi) in relation to its tensile cohesive strength. Thus, although the material itself can with-stand a tensile pull of 400 psi, only 150 psi of this capacity can be used before the sealant pulls loose from the substrate.

The silicone sealants have an acceptable peel adhesive strength. The peel strength value (20 psi) is good, but not outstanding.

The silicones, in flat sheet specimens, have an ultimate elongation of 450% which is roughly the same value reported for many urethanes and polysulfides. In practical applications, however, where the shape of the sealant is quite different, neither manufacturer would recommend a sili-cone for joint movements in excess of 50%.

Cyclic tension and compression is considered to be the norm of joint movement. The silicones, because of their high recovery, perform well in cyclic tension and compression testing. Let us assume a cycle of 50% ex-tension and 50% compression. The silicone sealant will be stressed when the specimen is extended. As the specimen is brought back to its initial width, the stresses in the specimen are relieved. This would not be true of a low recovery sealant. The low recovery sealant would relax inter-nally and, consequently, would have to be pushed back into the zero movement position. This same stress situation would repeat itself in the compression half of the cycle.

Because of the high recovery, the modulus of elasticity of silicones can be reported with excellent reproducibility of results. The plotting of the modulus curve of silicone is not as severely affected by the rate of test-ing as it is with the polymercaptans and polysulfides.

Silicone rubber, because of its stability and high recovery, would seem to be a logical candidate for preformed gaskets. There are some preformed silicone seals in use in building construction, and experimen-tal sections have been extruded for use in highway pavement joints. However, the price of gasket sealing methods, together with the noto-riously high price of silicones, has held their penetration of this market to a minimum.

The silicone sealants, along with many other elastomeric sealants, can be used as adhesives for specific applications. Table 8.1 lists quantitative

Table 8.1

Tensile strength	400	psi
Ultimate elongation ($\frac{1}{2}$ - × $\frac{1}{2}$ -in. specimen)	80–100	%
Tear resistance	40	lb
Adhesive strength	150	psi
Peel strength	20	lb/in.
Hardness, Shore A	35	
Recovery	97	%

Fig. 8.1 United Nations Plaza Apartments. Sealed with a one-component silicone sealant. (Courtesy of General Electric Company, Silicone Division)

Table 8.2   *Advantages and Disadvantages of Silicone Sealants*
(as reported by consumers)

Advantages	Disadvantages
1. One-component systems	1. Too expensive
2. Available in the widest range of colors	2. Require critical surface preparation *and* priming (adhesion highly dependent on both)
3. Color stability	3. Poor adhesion to concrete
4. High temperature resistance	4. Inferior elongation relative to polyurethanes and polysulfides
5. Excellent handling characteristics at ambient temperatures of − 35 F to + 140 F	5. Inferior tensile strength relative to polyurethanes
6. Good UV light and ozone resistance	6. Poor tear resistance
7. Nonsagging on vertical walls	7. Joint width limitations
8. Good adhesion to metal and glass	8. Shelf life too short
9. Exhibit no shrinkage	9. Require contact with atmospheric moisture to cure
10. Become tack-free in a short period of time (fast cure)	10. Cohesion too great for their adhesion
11. Excellent flexibility; permanently flexible	11. Pick up dirt
12. Almost 100% recovery from elongation or compression (very low compression set)	12. Unpleasant odor (release of acetic acid)
13. Good heat and chemical resistance	
14. Excellent durability or long life (about 30 years)	
15. Longer pot life (when mixed) than two-part polysulfides	
16. Good resistance to digging and gouging	
17. Nonstaining to most materials	

values for the physical properties of a typical single component silicone construction sealant.

8.4   *Application*

The silicone sealants must be applied into very clean joints. Cleaning of the joints is probably more critical with the silicones than with any other sealant. (See Section 6.5 on joint cleaning.)

Since the silicones in the uncured state are not soluble in water, solvents must be used for the applicator's clean-up at the end of the day. This contributes to the cost of the complete installation.

Although the silicones will adhere to some substrates without the use of a primer, priming is recommended for any working joint. Different primers are used for metal and concrete substrates.

The silicones are used for exterior joints in metal and concrete curtain wall panels, metal-to-glass sealing, control joints, expansion joints, and the sealing of hairline cracks for which the less workable sealants cannot be used. They are not generally used in nonworking joints where their properties are not needed. The United Nations Plaza Apartments, sealed with this material, is shown in Fig. 8.1.

8.5 Summary

The silicones are high quality, high recovery, elastomeric sealants. They are quite stable in both the uncured and cured state. They are easy to work with, and can be tooled to a very neat joint. They have very good color stability and will not stain most substrates. Also, they are best used in exposed locations where their properties can be utilized. Table 8.2 summarizes the advantages and disadvantages of the silicones as observed by the architects who specify these sealants and the caulking contractors who apply them.

9

Urethane Sealants °

9.1 Introduction

The urethane sealants are perhaps the most interesting new construction sealants introduced to the industry in recent years. The urethane polymer is extremely versatile and can be compounded for a wide variety of uses. Urethanes have been widely used as rubbers, coatings, foams, and fibers; and in 1960 the two-component urethane sealants were introduced to the construction market.

The properties of the urethanes, in general, are intermediate between the polysulfides and the silicones. They are considered to be a high recovery sealant with good workability and good adhesion.

The urethane sealants are available as both one- and two-component sealants. The two-component sealants are available in both self-leveling and nonsag grades, and the one-component is available in the nonsag grade. Three-component urethanes are produced, but these are specialty items and account for only a minor percentage of the urethane production.

The urethane sealants are quite versatile because the finished product can be formulated in a variety of ways

1. The basic urethane polymer can be prepared in different ways through the use of various organic constituents.

2. The urethanes can be, and are, often prepared as combination seal-

° The two names, "urethane" and "polyurethane," are both used when referring to this class of sealants. The two names are synonymous. The simpler name, "urethane," will be used throughout this book.

ants, such as urethane acrylic or urethane epoxy.

3. The urethanes can be blended and modified with fillers and plasticizers, as can other elastomers.

Thus, with these three compounding routes open, the urethanes not only offer the widest possible range of sealant properties, but also show up as potentially the lowest cost premium sealant on the market.

The early urethanes had two major defects—water sensitivity and color stability. Urethane foams are made by a urethane-water process and this same process took place to a minor degree when urethane sealants were exposed to moisture prior to complete cure: the sealants bubbled and caused irregularities or large voids in the sealant cross section. The second defect was a decided lack of color stability: the urethane sealants tended to turn amber in color after weather exposure. Both of these problems have been virtually eliminated by the major manufacturers, but the problems are of vintage recent enough for the image, in many instances, to remain. In the overall picture, these two problems must still be considered as minus factors by the consumer because they are problems that must yet be overcome by the smaller formulators and new formulators who enter the competition.

The urethane sealants are covered by all three of the major specifications for building work. TT-S-00230 covers the single component sealants and TT-S-00227 covers the two-component materials. The USASI 116.1 specification also covers the two-component sealants. For highway and airport construction, several urethanes extended with coal tar and other bituminous fillers have been introduced quite recently. These materials are intended to meet the requirements of Federal Specification SS-S-00200. This particular specification has, for some years, been accepted by the trade as a "polysulfide coal tar specification." The urethane sealants formulated for this specification are very low recovery materials and are not typical of the urethane sealant group. As with other elastomeric sealants, materials which meet the three building specifications contain approximately 40 to 50% polymer. The bitumen-extended sealants have a somewhat lower polymer content.

At this time there are three or four major suppliers of raw materials for urethane sealants and about ten formulators who compound the finished sealant. There are also two producers who manufacture their own basic polymer and compound the finished sealant. The number of formulators is expected to double within the next 5 to 10 years as the polysulfide formulators expand their product lines. This competition, together with the number of compounding options available, will probably lead to a general lack of uniformity in the finished sealant compounds. The existing specifications do cover the urethane sealants, but the segment of nonspecification sealing work is large enough to encourage the development of lower quality urethanes. Brand names and the manufacturer's reputation thus become increasingly important. Also, some industry groups, such as USASI and the Society of the Plastics Industry, are attempting to adopt a hallmark program which will serve to promote orderly growth.

9.2 Compounding

There are different routes in the preparation of the finished urethane sealant which can result in varying ratios of base component to curing agent for the two-component sealants. In general, the reaction results from an isocyanate component and a hydroxyl component combining to form urethane groups. The two components are sufficiently fluid to make for easy mixing. The ratio of components may vary from 1:1 to 10:1, depending on the formulation used. Compared to the polysulfides, the urethanes are somewhat easier to mix because the components are usually viscous liquids, whereas with the polysulfides the accelerator is usually a relatively thick brown paste. On the other hand, there is no diffusion cure in the urethanes and therefore they must be mixed thoroughly and uniformly, with a greater shearing action than is commonly used with the polysulfides.

It is difficult to tabulate any typical composition of a two-component urethane sealant. The one-component moisture-cured sealant will contain approximately 50% urethane prepolymer, 40 to 45% fillers, and 5 to 10% solvent to help the gunnability of the finished sealant. As with the polysulfides, proper manufacture of the single component urethane is more difficult and expensive than the two-component materials, and not all compounders are equipped for this type of production.

The fillers and extenders used in compounding urethanes are much the same as those for other elastomeric sealants. Carbon black is a good reinforcing filler as well as a coloring agent. Asbestos, ground silica, and the clays are also used. Pigments are used, but the range of available colors for the urethanes is not yet as broad as for the other sealants.

9.3 Properties

At the present time the urethanes are probably the subject of more intensive research than any other sealant. Urethanes are being used as building sealants, paving joint sealants, and, because of their immunity to biological degradation, they are widely used as seals for sewer pipe. Much of this field and laboratory data is available in previously published literature. The specific values for various properties which are reported here may be considered as average values for a high grade building sealant.

9.3.1   Odor

The urethane sealants do not have any significant odor problems. The single component material in the uncured state will have the odor characteristic of the solvent used in the compounding. As with any volatile solvent, adequate ventilation is recommended.

9.3.2   Toxicity

The urethanes are nontoxic and nonallergenic. They do require the use of a solvent for the operator's clean-up.

9.3.3   Solvent Resistance

The urethanes have very good oil resistance. They can be extended with coal tar to form low modulus paving joints sealants with resistance to jet fuel. The urethanes are not noted for good water immersion resistance. They must be applied to dry substrates, especially in the case of the one-component systems. Many of the urethanes are affected by continuous high humidity or water exposure.

9.3.4   Hardness

The urethanes can be formulated into low modulus, low recovery sealants with a Shore hardness as low as 5 to 10. A typical building sealant has a hardness value in the 20–35 range. Harder sealants are available for traffic-bearing joints.

The change in hardness with temperature of the urethane sealants is between the values indicated for silicones and polysulfides. Over a temperature spread of 140 F, the silicone changes in hardness by less than five points; the polysulfide varies by 30 points over the same range; the urethane sealant varies by approximately 15 points.

Parallel to hardness is the excellent abrasion resistance of the urethane sealants. They can stand scuffing and foot traffic quite well. This property is related to tear resistance as well. The urethanes are intermediate in tear resistance between the silicones and the polysulfides. The excellent abrasion resistance refers to the fact that the traffic-bearing urethane sealants are almost as hard as the silicones and have significantly better tear resistance.

9.3.5   Aging and Weathering

The urethane sealants have excellent resistance to ultraviolet exposure but only fair water immersion resistance. They have good resistance to ozone and are generally considered a very good weathering sealant. Most high quality urethane building sealants will survive for well over 1000 hr in the accelerated weathering chamber; they may therefore be used in exposed locations. These sealants have good recovery and show very little surface wrinkling in service. They also maintain elasticity over long periods of time without excessive chalking or crazing. Although the major manufacturers have apparently solved the color problem, long-term field verification of color stability must still be established.

9.3.6   Modulus of Elasticity

The modulus of elasticity of urethane sealants is usually somewhat higher than that of the polysulfides, thereby indicating a generally stiffer product. There are enough compounding routes open so that lower modulus urethanes can be developed by altering the structure of the polymer. The modulus can be lowered also by the addition of plasticizers, but some urethane formulators feel that this is not the best approach. These formulators believe that plasticizers may possibly lose their effectiveness over long time periods, so that the user, after a few years, winds up with the original high modulus sealant.

The modulus of the urethanes, because of their high recovery, is not affected too severely by changes in the rate of testing. A moderate testing rate of $\frac{1}{8}$ in./min is a good practical testing speed to use with the urethanes. Specimens of $\frac{1}{2} \times \frac{1}{2} \times 6$ in. tested at $\frac{1}{8}$ in./min, will yield a 50% modulus value of 60 to 100 psi. This value is low enough for the urethanes to be used safely with concrete substrates.

9.3.7   Ultimate Elongation

The ultimate elongation of the urethanes can probably be varied over a range wider than that available for any other sealant. Ultimate elongation is, of course, related to recovery. A piece of well-chewed bubble gum has an ultimate elongation in excess of 1000% but, since it has virtually no recovery, it would make a very poor sealant for working joints. Although the urethanes can be formulated into low recovery sealants, the normal urethane building sealants retain their high recovery characteristic over a broad range of ultimate elongation values. As further research develops the urethanes, this combination of high elongation with high recovery makes them look very promising. Sealants, formed into $\frac{1}{2} \times \frac{1}{2}$-in. specimens, for use in buildings, have elongation values of 80 to 100%.

9.3.8   Creep and Stress Relaxation

The urethane building sealants show very little creep or flow. Regarding recovery, the urethanes are better than the polysulfides, but not as good as the silicones. Specimens of urethane building sealant, extended and blocked into position for 24 hr, show almost complete recovery when released. The same specimens blocked into place for several months show a recovery of approximately 80%. Since recovery is almost inversely proportional to tear strength, the high recovery of the urethanes indicates a tear strength which is better than the silicones, but not as good as the polysulfides.

Table 9.1 is a summary of typical properties of a urethane building sealant. The values are based on $\frac{1}{2} \times \frac{1}{2}$-in. specimens.

Table 9.2 is a summary of the advantages and disadvantages of the urethane sealants as seen by the consumers.

Table 9.1

Tensile strength (cohesive)	300 psi
Adhesive strength	160 psi
Ultimate elongation	80–100 %
Tear resistance	70 lb
Hardness, Shore A	30

9.4   Application

The two-component urethane sealants are packaged in preproportioned kits for delivery to the job site. The two components are usually viscous liquids; therefore the curing agent can be easily poured into the base component. The colors of the components are usually different and thereby aid the mixing operator. A pot life of 5 to 6 hr is normal, but

Table 9.2   Advantages and Disadvantages of Urethane Sealants
(as reported by consumers)

Advantages	Disadvantages
1. Handle wide joints and large movement without failure	1. Not as good elasticity as the poly-sulfides
2. Excellent elongation and recovery	2. Poor color stability—they turn amber
3. Excellent resistance to penetration for use in horizontal traffic-bearing areas	3. Poor water immersion resistance
4. Can be formulated without the use of fillers to a greater extent than any other sealant can be formulated with the use of fillers	4. Absolutely cannot be applied to a wet joint—no adhesion
5. Tough; abrasion resistance	5. Some adhesion problems due to a high modulus; high cohesive strength, which can promote adhesion failures, especially in poorly coherent masonry structures
6. UV resistance	6. Too sensitive to moisture pickup (will tend to either set up in the can or gas-foam) and thus too sensitive to handle
7. Excellent adhesion to a wide variety of substrates (two-component system) with proper surface preparation and priming	7. Require significant surface preparation and priming to obtain adhesion
8. Higher tensile strength and hardness, and UV resistance than polysulfides	8. Poor package stability (shelf life too short)
9. Excellent cohesive strength	9. Poor flow control
10. Good durability or long life (20–30 years)	10. Weather conditions of more than 70% humidity can reduce their effectiveness
11. Fast cure (two-component system)	11. Two-component systems have the usual problems of mixing, but in addition they must be mixed completely, thoroughly, and uniformly (10 min by the clock) with more shearing action than is required by other sealants
12. Low cold flow	
13. Negligible shrinkage	
14. Resistance to hardness increase due to oxidation	
15. Excellent resistance to compression set	12. Cure rate for single component systems is slow and the elastomeric properties are not attained for a considerable period
16. Excellent resistance to notch propagation	
17. Excellent tear resistance	13. One-component systems have experienced internal foaming (cellular structure) of the applied bead of sealant
18. Broad range of cured hardnesses available	
19. Excellent chemical resistance	

this can be varied. Most two-component urethanes are mixed at the job site by using a blade agitator and a slow speed electric drill. Since there is no diffusion cure in the urethanes, mixing is extremely critical. The urethanes must be mixed slowly, thoroughly, and uniformly to assure a uniform finished product. This mixing, of course, adds to the labor cost.

The one-component urethane sealants are usually furnished to the job site in $\frac{1}{10}$-gal cartridges. For large-scale operations, the material is usually furnished in 5-gal pails, and bulk-loading caulking guns are used.

The viscosity or gunnability of the urethanes is approximately the same as for the polysulfides. The urethanes extrude well from standard caulking equipment and fill any voids or irregularities in the substrate.

The urethanes are not as sensitive as the polysulfides to temperature during cure. Although the cure is accelerated by heat and humidity and is retarded somewhat by colder temperatures, the urethanes can tolerate cold weather installation. On the other hand, these materials are very sensitive to water during the curing stage; hence they cannot be applied to damp substrates.

The urethanes require very clean joints in order to obtain proper adhesion. These sealants will adhere to metal substrates without the use of a primer, but every manufacturer recommends the use of primers with all types of substrates. The urethanes adhere well to glass; this property, together with their good UV resistance, makes them desirable for glazing work.

The urethane sealants at the present time are priced competitively with the polysulfides and are used in the same types of working joints. Because of their toughness and abrasion resistance, they are often used in joints in industrial floors and sidewalks. The Federal Mogul Building, shown in Fig. 9.1 is an example of a structure sealed with this material.

Fig. 9.1 Federal Mogul Building, Detroit, Michigan. Sealed with a urethane sealant. Courtesy of Giffels & Rossetti, Architects & Engineers)

9.5 Summary

The urethanes are competitively priced, high recovery, high quality sealants. Their cohesive strength is greater than their adhesive strength. The adhesive strength is still somewhat low for such a high recovery material. The tear resistance of the urethanes is excellent, and they can tolerate cyclic deformation of up to 50% for long periods without failure. They are available in both self-leveling and nonsag grades for the two-component materials, and in a nonsag grade for the one-component materials. They have good UV resistance, but only fair water immersion resistance. The color stability over extended time periods needs field verification. Mixing is especially critical. In addition, this sealant requires very clean joints and priming in order to function effectively.

The urethane sealant, because of its versatility, is the subject of more intensive research than any other sealant. The urethanes already have the lowest raw material cost of the premium sealants, and this intensive research should further lower the cost and upgrade the properties of this material. At present the urethanes occupy a significantly smaller segment of the construction market than the polysulfides, but a rapid growth in the use of urethanes is predicted as the price per performance picture changes.

10

Polymercaptan Sealants

10.1 Introduction

The polymercaptans are a new group of sealants just entering the sealant market. They are so new to the market that very little information is available about their formulations and properties. They are, as the name implies, mercaptan-terminated polymers. In this regard they are similar to the polysulfides. However, they also differ from the polysulfides: the backbone of the polymer chain is formed from some of the same raw materials used in the urethanes. This combination would naturally suggest that the polymercaptans, with respect to their properties, are intermediate between the polysulfides and urethanes. The attempt is made, in compounding the polymercaptans, to retain the better features of both the polysulfides and the urethanes.

The polymercaptan sealant is available in both one- and two-component forms and in both self-leveling and nonsag grades. The two-component sealant is cured by the addition of a curing agent which supplies oxygen. Lead dioxide is often used, as it is with the polysulfides, although other metallic oxides and organic oxidizing agents can be used. The two-component sealant has been available for a very few years, and the one-component material is currently making its entry into the sealant market.

The polymercaptan sealants are expected to meet the requirements of Federal Specifications TT-S-00227 and TT-S-00230. The specification material contains approximately 40% polymercaptan polymer.

Along with most other elastomeric sealants, the cure of the one-component sealants is achieved by atmospheric oxygen. Heat and humidity accelerate the cure and, conversely, a cold atmosphere retards it.

At this time, the polymercaptans do not have the versatility of either the polysulfides or the urethanes. The sealant polymer is available in a single grade which bears the trade name DPM 1002 °. Properties are varied by the addition of fillers and plasticizers.

The polymercaptans are another exception to the general roles of raw material supplier and formulator. The basic polymer and the finished sealant are both produced by Diamond Shamrock Corporation. This "in-house" formulation of the finished product results in excellent quality control of the finished product. Also, the manufacturers are well aware of being relative newcomers in a highly competitive market. They are also aware of the architect's natural reluctance to specify a product which has a very short history of field performance. All of these factors have combined to produce sealants which show excellent possibilities.

10.2 Compounding

Information concerning the compounding of the polymercaptans is as yet largely unavailable. The limited information which has been published shows that the polymercaptans are filled and plasticized in much the same fashion as the polysulfides. Curing can be accomplished by the addition of 7½ parts of lead dioxide paste per 100 parts of base resin. This is the same ratio used with the polysulfides and is somewhat too unbalanced for easy mixing.

Fillers used with the polymercaptan are carbon black, titanium dioxide, calcium carbonate, coal tar, and perhaps others. The carbon black and calcium carbonate increase modulus, hardness, and tensile strength. The coal tar extender is used in paving sealants and results in a low modulus product.

The polymercaptans are plasticized with much the same organic plasticizers as those used with the other elastomeric sealants. The plasticizers lower the modulus, increase the ultimate elongation, and, in some cases, give low temperature flexibility. As with the polysulfides, the plasticizer is often added to the curing paste at the factory to bulk up the curing agent component. This somewhat offsets the imbalance of polymer to curing agent and makes for easier field mixing.

10.3 Properties

Because of the short history of field applications of the polymercaptans, it is difficult to state definitive ranges of physical properties. The

° Trademark of Diamond Shamrock Corporation.

values listed here are based on the relatively limited published literature, the manufacturers' technical data sheets, and the author's own limited laboratory experience.

The two-component sealants are virtually 100% solids and thus there
is no shrinkage when the material cures. The one-component materials
probably have about 5 to 10% solvent added to improve gunnability because laboratory specimens of these materials indicate a slight amount
of shrinkage.

10.3.1   Odor

The polymercaptans have the hydrogen sulfide smell characteristic of
the polysulfides, but not to such a marked degree. The coal tar modified
sealants, of course, have the distinctive coal tar odor. The one-component materials have the odor associated with the solvent used in their
preparation.

10.3.2   Toxicity

The polymercaptans do not indicate any problems with toxicity or al-
lergies. They do require the use of a solvent for operator's clean-up.

10.3.3   Solvent Resistance

The solvent resistance of the polymercaptans has yet to be proved on
a long-term basis. The backbone of the polymer is similar to the urethanes, which are only fair in their resistance to water immersion. On
the other hand, these resins are similar to the polysulfides with respect
to mercaptan-terminated polymers. The polysulfides are noted for resistance to both water and oil. It remains to be seen how this class of materials will perform over a period of years. Laboratory tests indicate good
oil resistance and water immersion resistance sufficient to meet the requirements of the Federal specifications.

10.3.4   Hardness

The polymercaptan building sealants are available in a rather limited
range of hardness values. The hardness increases with fillers and decreases with the addition of plasticizers. Hardness values tested in the
laboratory indicate very good stability with changes in temperature.
The usual two-component sealant has a Shore hardness of 30–40 at 70 F
which makes the material very desirable for traffic-bearing joints. This

value of hardness, however, is somewhat high for use in building joints, especially those with concrete substrates. Since these are relatively new products, further research will undoubtedly extend the range of sealants available.

The polymercaptans have succeeded in combining a relatively high hardness value with a good tear strength; although no work has yet been done in this area, indications suggest very good abrasion resistance.

10.3.5 Aging and Weathering

The polymercaptans will successfully endure over 1000 hr in the accelerated weathering chamber, which indicates that they are candidates for exposed, one-stage sealing. The building sealants should show aging and weathering properties similar to the polysulfides and urethanes.

10.3.6 Modulus of Elasticity

The modulus of elasticity of the polymercaptans is comparable to that of the silicones and urethanes and generally somewhat higher than that of the polysulfides. The polymercaptans, because of the single grade of polymer, are somewhat limited in the modulus range available. The modulus can be somewhat varied by the addition of fillers and plasticizers. As with the other polymeric sealants, modulus is increased by fillers and decreased by plasticizers.

The polymercaptans are a moderate recovery-type sealant. Consequently, the modulus value is affected by the rate of testing in much the same manner as with the polysulfides. Specimens of polymercaptan, $\frac{1}{2} \times \frac{1}{2}$ in. in cross section, and tested at $\frac{1}{8}$ in./min, will yield a 50% modulus value of 60 to 100 psi. This value is at the upper end of the range of values which are acceptable for concrete substrates.

10.3.7 Creep and Stress Relaxation

The polymercaptan sealants are moderate recovery sealants. Their rheological properties are intermediate between the properties of the polysulfides and urethanes. The polymercaptans show less recovery than the urethanes, but about the same recovery as the higher modulus polysulfides. The creep curve for the polymercaptans is flatter than the curve normally associated with the polysulfides.

The polymercaptan sealants have a definite curve for stress relaxation: a curve which is not as steep as the polysulfide curve for short time values and which flattens out into a longer stress relaxation time.

The polymercaptans, when extended 50% and blocked into place for 24 hr, will recover approximately 60 to 80% upon release. This recovery value drops to very low values if the specimens are extended for several months.

For the user, this combination of properties indicates that the polymercaptan will relieve the peak stresses induced by "stick-slip" joint movement, with good shape retention. However, this same sealant, if held in a deformed position for extended periods, will flow and show little recovery.

10.4   Application

The two-component polymercaptans are packaged in preproportioned kits for delivery to the job site. The base component is a viscous liquid and the curing agent is usually a paste. A pot life of 1 to 2 hr is normal. Cure time for the two-component materials is 48 to 72 hr. Mixing of these components is similar to the mixing of other two-component sealants. The structure of the material would indicate that complete and thorough mixing is necessary to effect a cure.

The single component polymercaptans are packaged in $\frac{1}{10}$-gal cartridges and in 5-gal pails for bulk use.

The polymercaptans extrude well and are easy to tool into the joint.

Laboratory work indicates some increase in viscosity with cold temperature, which makes gunning the sealant at cold temperatures a little more difficult. There is at present very little feedback information to indicate how the material performs under variable construction site conditions.

At the job site, the polymercaptans are eligible for exterior use in the one-stage weatherproofing of buildings. Laboratory tests indicate good adhesion to metal, concrete, and glass. This property makes them logical candidates for the sealing of joints in precast concrete panels or porcelainized metal panels. They would also be suitable for panel-to-mullion joints, for coping joints, and for face glazing with metal sash. Their relatively high hardness value makes them look desirable for horizontal joints in sidewalks and roof decks.

The polymercaptans have not as yet made any significant penetration into the highway sealant market. However, the manufacturers are well aware of the volume of this market and are currently testing several products for this purpose.

10.5 Summary

At the time of this writing, there is not enough feedback from finished jobs to furnish a table of advantages and disadvantages as seen by the consumers.

11

Butyl Caulks *

11.1 Introduction

The growth of butyl rubber began in the years of World War II. The scarcity of natural rubber led to the construction of butyl rubber manufacturing plants by the U.S. Government. After the War, these facilities were purchased by private industry, and the basic technology of the manufacture of butyl rubber was established. In the 1950's, the first butyl rubber caulks appeared in the construction market. In new nonresidential construction, the butyls have been second only to the polysulfides in total volume since they were introduced.

The butyls are elastomeric sealants with good flexibility but low recovery, and they are very durable. They have a fairly long history in the nonresidential construction industry and quite recently have invaded the consumer market in force. Currently about 35 companies manufacture a butyl sealant. In some of the European countries, where the butyls are better known and more widely available than other elastomeric sealants, they have captured a large share of the total construction market.

It is quite significant that the growth of two-stage weatherproofing in Europe is related to the growth of butyl sealants. In Europe, the butyl caulks have been widely available, and these materials ideally suit the requirements of the wind seal in two-stage weatherproofing. This is not to say that the butyls caused the rapid growth of two-stage weatherproofing in Europe; for two-stage sealing simply makes such good sense

---

* The terms "butyl caulk" and "butyl sealant" are used interchangeably throughout the construction industry.

that it needs no further justification. However, the two growth trends are definitely related. Also, as this method of sealing gains wider acceptance in the United States, the butyls will probably offer stiffer competition to the polysulfides, urethanes, and silicones in the new nonresidential construction market.

The butyls are compounded by a wide range of formulators, from the largest manufacturers down to very small plants. Very few would be equipped to manufacture the basic butyl rubber. This basic ingredient is furnished by one or two major suppliers. In addition to the normal compounding, in which the formulators purchase raw materials and blend a sealant, a fairly large amount of butyl sealant is manufactured by one of the larger compounders and sold to smaller formulators or paint companies who market it under their own private label. Consequently, while there are about 35 to 40 formulators of butyl sealants, there are as many as 50 to 60 separate brand names available on the market. This, of course, contributes to a wide variation in the quality of butyl caulks which are available.

The formulator with experience in blending butyl sealants has placed himself in a good position with respect to the "do-it-yourself" market among the homeowners. The homeowner likes "gimmicks" and will buy specialty items if he has a problem. The butyl is versatile enough in its properties for it to be packaged and sold satisfactorily as "Gutter and Downspout Sealer," "Driveway Crack Sealer," "Concrete Block Sealer," and as other specialty uses. These specialized items can be expected to proliferate.

11.2 Compounding

Two basic approaches to the compounding of a butyl caulk lead to different types of sealants. Formulation using a gum solution of unvulcanized rubber results in a product, generally for the light construction or consumer market, which meets or exceeds the requirements of Federal Specification TT-C-598. This material competes mainly with oil-based caulks and latex caulks. Formulation using a vulcanized butyl rubber results in a higher performance (and price) sealant. Some of the sealants formulated in this fashion have reportedly met the requirements of Federal Specification TT-S-00230. These materials represent competition for the elastomeric sealants.

The butyls, in general, cure by release of a solvent. Although tacky, they have little natural adhesion; hence adhesion additives are blended into the formulation so that the finished sealants have good adhesion to most substrates without the use of a primer.

The butyls will accept fairly high filler loadings. Calcium carbonate is the most common filler, although asbestos fibers, fibrous talc, and coloring pigments are also added. A typical formulation might include 20% polymer and 50 to 55% filler. As with other elastomeric sealants, fillers add to the strength of the sealant and decrease the ultimate elongation.

Plasticizers are a more serious consideration in compounding butyls than they are in compounding other sealants. The butyls cure by solvent release and therefore the solvent might be considered as a temporary plasticizer. It makes the mix plastic or extrudable enough so that it can be placed in the joint, but it also evaporates on cure, with the resultant shrinkage. Other "permanent" plasticizers can be added to give more tackiness or better low temperature flexibility. Some of the plasticizers which are common to other elastomers, such as oils and Aroclor *, are used. Also used very commonly is the petroleum derivative, polybutene. Polybutene has high tack, and mixtures of this additive in the sealant make it more sticky, or tacky. However, too much of this plasticizer results in a sealant which remains sticky and tends to pick up surface dirt. So the trick in butyl formulation is to add enough of the permanent plasticizer to give good flexibility, but not so much as to make the sealant permanently sticky. The formulator must also add enough temporary plasticizer (solvent) to make the sealant readily gunnable, but not so much as to give excessive shrinkage after the sealant is placed in the joint.

The good quality butyls also contain small quantities of a drying agent to help the sealant skin over after it is placed in the structure.

11.3   Properties

The butyl caulks are one-component, gun grade sealant materials. Depending on formulation, the tack-free time may be as long as 6 hr and complete cure through of the sealant may take a month or more. The butyls are relatively soft, low modulus materials with low recovery. They have very good shelf life and are available in white, black, aluminum, and tan.

11.3.1   Hardness

The butyls are relatively soft materials which have a Shore hardness of about 15 for the unvulcanized butyls and 20 to 30 for the vulcanized

* Trademark of Monsanto Chemical Company.

compounds when fully cured. The unvulcanized sealants are soft and compliant, and thus they can easily be picked or gouged from the joint by vandals. The higher grade butyls are not so susceptible to this type of damage.

The butyls will change in hardness when exposed to cold temperatures. The change in hardness is about 20 points, which is comparable to the urethanes.

11.3.2   Toxicity

The butyls have no toxic or allergenic problems. They do require the use of a solvent for the operator's clean-up. Odor is also no special problem. The sealants have the odor which is characteristic of the solvent being used. The usual precautions regarding ventilation and skin care associated with any solvent apply to the butyls.

11.3.3   Modulus of Elasticity

The modulus of the butyl caulks is, of course, dependent on the type of sealant being formulated and tested. The vulcanized butyls have a higher modulus. The modulus is affected by testing rate since the butyls are relatively low recovery materials. A vulcanized butyl in the $\frac{1}{2}$-× $\frac{1}{2}$-in. specimen size yields a 50% secant modulus of about 50 to 80 psi when tested at $\frac{1}{8}$ in. / min. This is a relatively low value of modulus and makes the butyls suitable for use with almost any type of substrate.

11.3.4   Ultimate Elongation

The butyl sealants do not have the elongation that might be expected of such a low modulus material. In general, as modulus goes up, elongation goes down, and vice versa. However, the butyls in the $\frac{1}{2}$- × $\frac{1}{2}$-in. specimen size have an ultimate elongation of only about 60 to 70%. This is no special problem since the manufacturers recommend the butyls for use in moderate movement joints (not over 20%).

11.3.5   Aging and Weathering

The butyl sealants age and weather very well without cracking or surface crazing. They have good UV resistance and, consequently, are often used in exterior glazing work. A well compounded butyl will withstand 1000 hr in an accelerated weathering chamber; hence it can be used for exterior, one-stage sealing.

When exposed to extended aging and weathering, the butyls are somewhat unique. Higgins [10] points out that when the butyls are aged for long periods, any oxidative change that takes place tends to break the polymer chains rather than cause a further cross-linking or stiffening. Thus, although the butyls are quite stable in the presence of oxygen, any attack that does take place will soften, rather than harden, the sealant. Consequently, these sealants retain their flexibility for very long periods of time.

11.3.6 Solvent Resistance

The butyls, when fully cured, have good resistance to oils and solvents. They also have good resistance to both acids and alkalis, and good water immersion resistance. Testimony to the water immersion resistance of these sealants is their growing market among boat owners.

11.3.7 Creep and Stress Relaxation

The butyls are a low recovery sealant. Their creep and stress relaxation properties are quite similar to those of a soft polysulfide. Specimens $\frac{1}{2} \times \frac{1}{2}$ in. in size, when compressed 50% and blocked into place for 24 hr, will show a recovery of about 20%. The same specimens blocked into place for several weeks indicate a minimal amount of recovery, perhaps 5 to 10%.

The butyl caulks will cycle well in tension and compression if the limits of movement are kept at 25% or below. If the cycling takes place at very low speed, say $\frac{1}{8}$ in. / hr, the specimens show some surface wrinkling. This would be a disadvantage in exposed one-stage weatherproofing, but should be no problem if two-stage weatherproofing is used.

The tear strength of the butyls is at the lower end of the range of values indicated for elastomeric sealants (40 to 80 lb), but, since these are deformable sealants, the tear strength is adequate for most applications.

The adhesive and cohesive strengths of the butyl sealants are fairly well balanced. The adhesive strength is generally slightly higher than the cohesive strength; therefore laboratory specimens generally fail by tearing of the sealant material rather than by its pulling loose from the substrate. At cold temperatures, however, the adhesive strength is somewhat lowered. This is a disadvantage in exposed locations, because cold weather contraction of the components of the structure causes the joints to widen.

Table 11.1 shows the advantages and disadvantages of the better quality butyl caulks as seen by the consumers.

Table 11.1 Advantages and Disadvantages of the Better Grade Butyl Caulks

(as reported by consumers)

Advantages	Disadvantages
1. Reasonable cost	1. Slow cure
2. Availability	2. High shrinkage
3. Good flexibility	3. High amount of compression set
4. Good adhesion to most substrates	4. Can only be used in joints with modest amounts of movement
5. One component (no mixing)	5. Ultimate elongation is low for such a low modulus material
6. Little surface preparation is required	
7. Very good water immersion resistance	

11.4 Application

The butyl caulks may be delivered to the job site in $\frac{1}{10}$-gal cartridges or in 5-gal pails. For the consumer market, they are available in cartridge form or an all-plastic squeeze tube.

The butyls may be safely applied over a temperature range of 40 to 120 F. Below 40 F they are difficult to extrude. Cold weather may retard the cure, but this is irrelevant since the cure of the butyls is already the longest of any commercial sealant.

The butyls require joint cleaning, but not to the critical degree demanded by some other elastomeric sealants. Laitance and loose concrete should be removed from the faces of concrete joints. Metal surfaces may be wiped with an oil-free solvent. Wood surfaces generally require only wiping with a clean cloth or soft brush. Glass should be wiped clean and dry with a soft cloth.

The butyl sealants do not require a primer with any type of substrate.

The butyl sealants may be used in exterior joints in curtain walls (vulcanized caulks) and in interior protected locations such as the air seal in two-stage weatherproofing. Both vulcanized and unvulcanized butyls have been used for two-stage sealing.

In addition to curtain wall joints, the butyls have found use in exterior metal sash glazing, reglets and cant strips for roofing, coping joints, door and window perimeters, and bedding joints. They have been widely used on the mating surfaces of heating and air-conditioning ducts, and the sealing of openings where pipes and ducts pass through roofs and partitions. They have also been used for the waterproofing of exposed screws and bolts. Because of the price differential, the butyls are not likely to lose the noncritical joint areas, such as duct sealing, even if the elastomerics continue the take-over of the curtain wall sealing market.

In the light construction and homeowner markets, the butyls should find increasing use in the lap joints of sheet siding, glazing, door and window perimeters, caulking of exterior and interior sills, and thresholds.

The homeowner uses the butyl caulks as a high performance oil-based caulk for glazing, and for door and window perimeters. He also uses the butyls for tub and shower caulking, flashing and roofing repairs, and general crack sealing.

The butyls have found an additional market with the installers of storm doors, storm windows, and protective siding. The butyls adhere well to aluminum, steel, and vinyl siding.

These butyl sealants are not intended for use in traffic-bearing joints, such as floors and sidewalks. In these applications, the butyls would be subject to penetration by dirt, stones, and spike heels.

11.5 Summary

The butyl caulks are single component, elastromeric materials which cure by solvent release. They are approximately 80 to 85% solids, which indicates some shrinkage of the sealant in the joint. They are a low modulus, low recovery sealant. The butyls come in two types: a vulcanized high performance sealant and an unvulcanized lower performance grade.

The butyls compete with the polysulfides, urethanes, and silicones in the high-rise construction market; in the consumer market they compete with the latex caulks and oil-based caulks. Their overall performance is considered to be intermediate between the polysulfides and the oil-based caulks, and they have a price tag to match.

The advantages of the butyl caulk are its low modulus, fairly good adhesion, and excellent water immersion resistance. The disadvantages of the butyl caulk are its high shrinkage, low recovery, and exceptionally long cure time.

Since the butyls are already well established in the construction industry, they are in a good position to compete in both the residential and nonresidential construction markets. The overall projection is for the butyl caulks to grow at a pace of approximately twice the overall sealants market.

12

Latex Caulks

12.1 Introduction

In many instances the growth of certain industrial products is related to the growth of other materials. This has been the case with the latex caulks. The growth of these caulks has definitely been tied to the growth of the latex paint market. As recently as 1956, oil-based paints were virtually unchallenged in the exterior paint market. However, by 1966, latex paints had taken over 40% of the exterior paint market and 80% of the interior market. The tie-in between the latex paints and the latex caulks is obvious. Whereas the oil-based caulking compounds will often bleed through a latex paint, leaving an unsightly color streak, the latex caulks are compatible with the latex paint system and produce a smooth looking job.

The name "latex caulks" covers many variations, but, in general, these materials are based on a vinyl acrylic or a polyvinyl acetate; both of which make a very good sealant. The vinyl acrylic is quite similar to the acrylic latex used in the paint formulations; and the polyvinyl acetate is similar in structure to the "white glues" sold as household adhesives. Both have very definite advantages.

The latex caulks are one-component, gun grade sealants. They are intended for use in joints with only moderate movement (up to 20%) and have good permanent flexibility, but little recovery. They offer the very definite advantage of quick clean-up. Spills and spots can be wiped clean with a damp cloth, and the applicator needs only water to clean his hands and tools.

The latex caulks are used mostly in the light construction market, rather than in heavy construction or high-rise buildings. This means that these caulks are used in residential work, schools, churches, light industrial buildings, motels and small office buildings, and apartments. These materials are purchased generally by either the homeowner or the painting contractor. The painting contractor is the prime buying influence. This buying pattern has been both cause and effect to the rapid growth of the latex caulks. Almost 90% of the painting contractors employ less than 20 men, and almost all painting contractors do some residential work. Also, almost 85% of the exterior residential paint market is for repainting, which demands caulking.

Since the bulk of these materials are purchased by the small painting contractor at the local hardware or paint store, the major sealant manufacturers have had to make a shift in order to get in tune with this new market. The major manufacturers were accustomed to dealing with architects and furnishing technical service to architects, engineers, and specialty caulking contractors; they were not geared for promotion at the local level. This has led to the growth of small, high quality formulators who have captured a large share of this market. Since the homeowners and small painting contractors purchase by brand name more than 50% of the time, the smaller formulators who have developed a reputation for quality should increase, rather than lose, their share of this growing market. Also, the big manufacturers were reluctant to produce a competitive oil-based caulk at about 20 to 40 cents per cartridge, but are quite willing to produce a higher grade caulk for the light construction market at about $1.50 per cartridge. Currently, about 20 formulators produce a latex caulk, and predictions speculate that by 1970 about 80% of all sealant manufacturers will include in their product line either a latex caulk or a solvent-based acrylic.

Another factor that has helped the growth of the latex caulks is the dealers. These local dealers realize there is little profit margin and a high rate of customer dissatisfaction with the cheaper grades of oil-based caulks; hence the dealers are also pushing the sales of higher grade caulks such as latex caulks, butyls, higher grade oil-based caulks, and silicones.

12.2 Properties

The latex caulks are one-component gun grade materials and are usually nonstaining to any type of substrate. They are generally tack-free in 1 hr and thus can be painted over almost immediately. They are normally about 80% solids, so there is some shrinkage when the sealant is placed

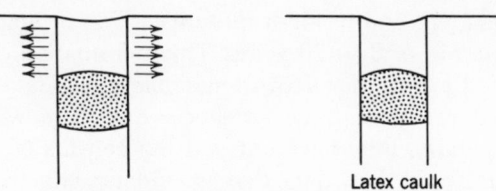

Latex caulk

Fig. 12.1 Comparison of sealant beads.

in the joint. The claim is sometimes made that this shrinkage sets up internal stresses in the body of the sealant even before the sealant is subjected to joint movement. This would be quite true if the latex caulks were highly elastic, high recovery materials. However, the stress relaxation curve for these materials is initially quite steep, so that the material has relaxed into equilibrium by the time that the sealant bead has cured through. As a matter of fact, the shrinkage may actually be somewhat beneficial in terms of stresses. Figure 3.5 showed that a tooled joint tends to produce more or less uniform stresses on the substrate as the joint moves. This figure is repeated here, along with the cross section of a latex caulk (Fig. 12.1). The similarity in the shape factors of these two sealant beads is quite obvious.

The latex caulks are relatively low modulus materials with only a slight amount of recovery. They have an ultimate elongation of about 60% when formed into $\frac{1}{2} \times \frac{1}{2}$-in. specimens. They have a good balance of adhesive and cohesive strength and will generally fail at about 50 psi in the $\frac{1}{2} \times \frac{1}{2}$-in. size. These materials cycle fairly well in tension and compression, provided the movements are less than 25%. They are deformable sealants, which means they relax stresses quite rapidly and have little recovery. Specimens blocked into place at 50% compression show about 20 to 25% recovery after 24 hr. If blocked into place for several weeks, the recovery is virtually nil.

The latex caulks bond well to concrete, aluminum, brick masonry, and wood without the use of a primer. They require joint cleaning, but not to the same extent as the highly elastomeric sealants. The latex caulks may be applied to damp, but not wet, absorptive surfaces such as concrete, brick masonry, and wood.

These caulks have good UV resistance, but only fair solvent resistance. They can tolerate a great deal of intermittent water exposure, but not water immersion. They may be used in exterior joints exposed to rain, but not in swimming pool or pool deck joints.

These caulks have good color retention, but, because of the type of market for which they are formulated, they have generally been available in only white or gray. As the market grows, many of the colors currently available in latex paints will develop in the latex caulks.

The latex caulks age and weather quite well over a range of -20 F to $+150$ F, for a period of 8 to 10 years. This information comes as feedback from finished jobs rather than as manufacturers' data. These caulks are not intended to be high performance, one-stage weatherproofing sealants. Consequently, they have not had the benefits of the high powered research and published data that is evident with the polysulfides, urethanes, and silicones. As the larger manufacturers begin to divert research money into this profitable market, even better latex caulks will develop. Table 12.1 shows the advantages and disadvantages of the latex caulks as seen by the consumers.

Table 12.1 Advantages and Disadvantages of Latex Caulks
(as reported by consumers)

Advantages	Disadvantages
1. Fast skinning and cure rate	1. High degree of shrinkage
2. Immediate paint-over with the inherent labor cost savings	2. Poor water immersion resistance
3. Some limited flexibility	3. Poor resistance to penetration
4. Do not bleed through any paint	4. Poor low temperature flexibility
5. Nonstaining to most materials	5. Poor low temperature cure (not below 40 F)
6. One-component systems	6. Need to be painted in exterior applications
7. Good adhesion	7. Can withstand only modest joint movement at best, or will pull away from the joint surface
8. Good UV resistance	
9. Easy application and clean-up	
10. Good package stability	
11. No primers required	

12.3 Application

The latex caulks are available in $\frac{1}{10}$-gal cartridges and in a new all-plastic squeeze tube. They may be ordered in larger containers for quantity use.

The latex caulks may be applied at temperatures above 40 F. Below this temperature, the sealant stiffens and is difficult to extrude from the caulking gun. Since this sealant is water-based, it must be protected from freezing before cure.

The latex caulks are ideally suited for small joints ($\frac{1}{4}$ to $\frac{3}{8}$ in.) above grade and for door and window perimeters (Fig. 12.2), flashing, and glazing. The homeowner now realizes the superiority of this type of caulk over the oil-based caulk for such locations as flashing, downspouts, baseboards, sealing around air-conditioners, tub and shower

Fig. 12.2 Latex caulk sealing the joint between brick wall and wood window frame. (Courtesy of Sika Chemical Company)

caulking, and other locations which will probably not be painted for some years. The homeowner also uses these materials for sealing cracks in concrete, and concrete block, and for roof patching.

The small painting contractor uses these caulks in conjunction with latex paints. Currently, about 90% of the painting contractors still use the oil-based caulks in most areas, but this percentage is declining. Also, the recent appearance of a Federal specification for latex exterior paint has opened the door to a new market outlet.

The market for the PVA-based latex caulks should expand even faster as architects and contractors begin to realize what many homeowners already know—that the latex caulk also serves well as an adhesive. Homeowners use these materials for sticking a loose wall or ceiling tile back into place, wood gluing, and other miscellaneous adhesive applications. Their fairly good adhesive strength, their nonstaining property, and the advantage of quick clean-up with a damp cloth make the latex caulks ideal for these uses.

The latex caulks are not intended for exterior, one-stage weatherproofing in high-rise office and apartment buildings. They do not have sufficient adhesive or cohesive strength for these critical applications. Their low recovery and limited color availability also makes them questionable for this application. Their relatively high shrinkage may cause an appearance undesirable for these applications. They also do not indicate the weathering capabilities over extended periods (15 to 20 years) that might be required in these structures.

However, with the future growth of two-stage weather sealing, in which the sealant undergoes moderate movements in a relatively protected environment, the latex caulks, as well as the butyls, may present a strong challenge to some of the higher priced, higher performance sealants.

12.4 Summary

The latex caulks are basically designed for light construction, which is mostly residential; in this respect, they represent an upgraded caulking compound to compete for the market now largely dominated by the oil-based caulks. They offer the advantages of being a one-component material, and providing quick application and easy clean-up. They skin over rapidly, remain flexible, and can tolerate moderate joint movements. They are compatible with, and will not bleed through, latex paints.

The disadvantages are low strength (both adhesive and cohesive) and low recovery. They have a high degree of shrinkage and poor water immersion resistance.

The overall growth picture for the latex caulks is very bright. They should grow at a pace much faster than the overall sealants market because of both the entry of the larger manufacturers into the picture and also the growth of interior and exterior latex paint.

13

Solvent-Based Acrylic Sealants

13.1 Introduction

The solvent-based acrylic sealants were first introduced into the construction industry in about 1960. Since their introduction they have been used to seal many important structures and have established a good performance record. These materials are used mainly in the big construction field and are not particularly well suited for use by the homeowner or small contractor.

These sealants are single component, gun grade sealants that have excellent adhesion without the use of a primer. They are low recovery, elastomeric sealants that will self-heal in the event of a cohesive failure. The excellent adhesion and the self-healing feature make the solvent-based acrylics one of the probable prime beneficiaries of the growth of two-stage weatherproofing in the United States.

13.2 Compounding

The solvent-based acrylics are thermoplastic sealants prepared from monomers such as ethyl acrylate. Until quite recently, only two formulators were producing this type of sealant and thus information on compounding is relatively sparse.

The sealants are compounded from base polymers supplied by the large chemical companies. The acrylic sealants contain approximately 50% polymer. Many of the same fillers, such as calcium carbonate and asbestos, that are used with other elastomers can be successfully incorporated into the acrylics. As with other sealants, the asbestos fibers help to produce a thixotropic (nonsag) sealant and other fillers, in general, increase modulus and hardness.

The acrylic sealants can be produced in a wide range of colors. The color stability of the sealants is also very good.

These sealants cure by solvent release. They are generally 80 to 85% solids; hence there is some shrinkage after the sealant is placed in the joint. The solvent-based acrylics apparently have a very slow rate of solvent release. The skinning and cure rates are both quite slow. The surface of the sealant remains tacky for a day or so and the complete cure through of the average size sealant bead may take from two to five weeks.

13.3 Properties

The solvent-based acrylics are nontoxic and nonallergenic. They are a solvent-based sealant, however, and should be used with a normal amount of care. A solvent is required for operator's clean-up.

Odor is a distinct problem with the solvent-based acrylics. The odor is not only offensive in itself, but can contaminate food products if the sealant is used in food storage or preparation areas. Adequate ventilation is a must when using this type of sealant for interior applications.

The solvent-based acrylics have good resistance to a fairly broad range of chemical solvents, after the material has cured. The resistance to water is good, but not outstanding. These sealants can be used for exterior one-stage weatherproofing. They will tolerate a good deal of intermittent water exposure, such as rain flowing over the joints in a vertical wall. These sealants should not be used, however, for joints which are immersed in water such as in swimming pools. These sealants also show good resistance to salt spray. In addition, the UV resistance of the solvent-based acrylics is excellent, which makes them very good glazing sealants.

The solvent-based acrylics age and weather very well. They retain colors quite well for many years without change. Also, they show very little chalking or crazing of the sealant surface when exposed to the weather for several years. The solvent-based acrylics are formulated to meet the requirements of Federal Specification TT-S-00230.

The physical properties of the solvent-based acrylics are seldom tabulated because this sealant is such a low recovery material. These sealants are rubbery: for example, a specimen held in the hands will, if stretched, show some recovery. Actually, they are not true elastomers.

The stress relaxation is so rapid that it is virtually impossible to plot a modulus curve for this type of sealant. The specimen in the testing machine will stretch with virtually no change in load. The specimens, however, will stretch from 200 to 250% before failure. When failure does occur, it is almost invariably a cohesive failure. Tensile adhesion values are not reported because adhesive failure is almost unknown in a tensile test. The peel adhesion strength of the solvent-based acrylics is about 12 to 15 lb./in., which is quite good. For the consumer, this combination of properties means a sealant with excellent adhesion. However, this is accompanied by poor recovery; therefore this sealant is best suited for smaller joints and joints with limited movement (if exposed). In wider joints, the sealants will tend to wrinkle under cyclic joint movements. Another outstanding feature of the solvent-based acrylic sealants is that these materials will self-heal. When a cohesive failure occurs, if the two fractured portions of the sealant are pushed back together, the material knits together and will continue to function.

The solvent-based acrylics when fully cured have a Shore hardness of 20 to 25. This value increases by approximately 15 to 20 points as the temperature drops below zero F. This value of hardness is high enough to discourage prying fingers. However, because of the low recovery, this type of sealant should not be used in traffic-bearing joints. Table 13.1 lists the properties of the solvent-based acrylics as seen by the consumer.

13.4   Application

The excellent adhesion of the solvent-based acrylics makes them one of the most versatile sealants for building construction work. They can be used for glazing, control joints, glass-to-mullion joints, and the pointing of brick and stone masonry. When applied into exterior sight-exposed joints, the solvent-based acrylics should be used with the proper shape factor. However, these sealants require no priming and can be successfully used in joints which have had only a minimum amount of cleaning. The solvent-based acrylics do not require the "hospital clean" joints demanded by some of the more elastic sealants. Because of their high adhesion and low recovery, these materials can be used for corner beads, reglets, and oddly shaped joint openings that many of the other elastomers cannot tolerate. This versatility and ability to tolerate some abuse are two of the prime reasons these materials are selected by architects. The architect can select *one* sealant for the building which will function in exterior panel joints, and also as heel filler beads for glazing,

Table 13.1   Advantages and Disadvantages of Solvent-Based Acrylics
(as reported by consumers)

Advantages	Disadvantages
1. Require no priming	1. Exhibit relatively poor low temperature elasticity
2. Require only minimum surface preparation	2. Can only be used in small width joints
3. Good elongation and excellent adhesion	3. Poor shear strength
4. Good color stability and available in many colors	4. Poor recovery (high compression set)
5. Highly resistant to digging or gouging	5. Must be heated to about 120 F to be applied
6. One-component systems	6. Slow skinning and cure rate
7. Excellent UV resistance	7. Remain slightly tacky when some pressure is applied and thus tend to pick up dirt and dust under these conditions, requiring a slicking agent to seal the surface and reduce dirt pickup
8. Self-healing (if they break cohesively because of large joint movement, they will rejoin or self-heal)	
9. Good durability or long life (about 20 years)	8. Relatively poor flexibility
10. Good moisture and chemical resistance	9. Have a strong odor which is offensive and can contaminate food products
11. Remain elastic indefinitely	10. Poor water immersion resistance
12. Nonstaining to most materials	11. High cold flow
13. Negligible shrinkage	12. Poor penetration resistance (cannot be used in traffic-bearing areas)
14. Withstand dynamic movement	

sealing the perimeters of air-conditioning ducts, and so forth. The Toronto City Hall, shown in Fig. 13.1, is an example of a structure sealed with this material.

The solvent-based acrylics do have disadvantages. The slow cure makes them unsuitable for the homeowner or small contractor. If used in exterior sight-exposed locations, they should be confined to narrower joints or joints with a limited amount of movement. If used in wide, sight-exposed joints the low recovery will cause the material to wrinkle badly and show a very unsightly appearance. Because of the slow cure rate, a slicking agent should be used over the surface of the joints after they are installed to prevent discoloration by dirt pick-up. The solvent-based acrylics are thermoplastic materials and must be gently warmed before they can be extruded from the caulking gun. This need for heaters is an added cost and nuisance for the contractor. Commercial heating chests are available, but some contractors have made their own by

Fig. 13.1 Toronto City Hall. Sealed with a solvent-based acrylic sealant. (Courtesy of Tremco Manufacturing Company)

mounting a few light bulbs inside a picnic chest. Newer materials which are gunnable at 70 F are currently being advertised; however, long term field verification is needed to determine whether these newer materials are actually an improvement over the older solvent-based acrylics which had established an excellent nine-year service record in many structures.

14

Oil-Based Caulks

14.1   Introduction

The oil-based caulking compounds were the first sealants to be used commercially and still account for approximately 55% of the total sealants market. Before the introduction of the elastomerics in the 1950's, the oil-based caulks were virtually unchallenged and had practically monopolized the market. One of the original specifications for a putty or sealant consisted of simply 88% whiting (calcium carbonate) and 12% linseed oil. Since the introduction of the elastomeric sealants, the oil-based caulks have lost their major position in the larger building construction segment of the market but they still dominate in residential construction, light industrial construction, maintenance, remodeling, and repair. Despite the loss of part of the market to the elastomerics, the oil-based caulks are not quite ready to "roll over and play dead." Bieneman [11] has shown that there is some high grade research being conducted on the properties of these materials. The benefits of this research have shown up in higher grade caulks already on the market, and further improvements should follow shortly. In the overall picture, the oil-based caulks will continue to lose ground in the marketplace, but not at the pace of the last 15 years. The losses will be mostly to the butyls and latex caulks which are gaining ground in residential work, repair, and remodeling. However, the oil-based caulks, in their present and improved forms, will be a major factor in the sealant market for many years to come.

The oil-based caulks are available in a variety of formulations. They may be made from a number of oils such as linseed oil, fish oil, soybean oil, tung oil, castor oil, and many others. Some of these are drying oils, and some are not. Some have better pigment wetting properties and some tend to bleed out of the caulk, thereby staining porous substrates. The oils may be blended to secure better properties and may also be combined with organic materials, such as polybutene, to provide more tackiness to the sealant. Research is now beginning to show which oils supply which properties and what types of filler loadings are needed for the best caulk. To date, research indicates that blown soybean oil is the basic ingredient of the higher grade caulks.

At the job site, the contractor wants a material that is easy to work with, skins over quickly, and remains somewhat flexible throughout the mass of the sealant bead. The material that meets these requirements can be painted over very soon after installation and should have a life expectancy of 5 to 10 years.

The oil-based caulks are not intended to be elastic materials and should not be used in working joints. However, the better grade caulks will accommodate movements of up to 15 to 20%. These sealants have virtually no recovery, but a good skin-forming caulk will tend to self-heal after a cohesive rupture.

There are actually four Federal specifications which cover the oil-based caulking compounds. Specification TT-C-598 is for a "Caulking Compound, Oil and Resin Base Type (for Masonry and Other Structures)." Specification TT-G-00410c covers a "Glazing Compound, Sash (Metal) for Bedding and Face Glazing." Specification TT-P-791a states the requirements for a "Putty, Pure Linseed-Oil for Wood Sash Glazing." Specification TT-P-781a covers a "Putty and Elastic Compound for Glazing Metal Sash." Of these four specifications, three are strictly for glazing use, and only the 598 specification covers a multiuse compound. Most of the higher grade oil-based caulks will meet the requirements of this specification.

14.2 Compounding

Compared to the elastomerics, the oil-based sealants are relatively simple substances. They consist of an oil, a bulking filler, and additives to impart special properties. The formulation of a good caulk, however, is not a simple job. The chemically curing sealants go through a definite chemical reaction process which is completed in a relatively short time, resulting in the cured sealant. The oil-based caulk ideally should skin over quickly and then the reaction of cure should stop or slow down dramatically so that the bulk of the sealant mass remains flexible for a period of years. It takes a high degree of competence by the formulator to select the proper oil, the amount and gradation of filler, and the desired additives in order to produce the best finished product.

Since there is such a wide variety of oils being used, the properties of the finished product vary over a wide range. Giordano [12] has broken these materials down into the following subgroups for easy reference:

Class I —Rapid hardening putties
Class II —Plastic glazing compounds
Class III—Skin-forming plastic joint sealants

Since this classification has been presented publicly at an ASTM conference, it will be convenient to follow it.

14.2.1   Class I—Rapid Hardening Putties

The materials in this group are commonly based on a drying oil, such as linseed oil, and a calcium carbonate filler. These materials are suitable for interior glazing work and meet the requirements of Federal Specification TT-P-791a. These are among the cheapest of the oil-based materials. They harden gradually and must be painted regularly. These are generally knife grade materials. Their consistency makes them easy to work with and many glaziers are therefore reluctant to change to other, softer materials. These materials are more nearly fillers than sealants because they accommodate very little, if any, movement.

14.2.2   Class II—Plastic Glazing Compounds

These caulks contain some drying oil and some other types of oils or additives to give more plasticity. Castor oil and some fish oils can be used for this purpose. In lieu of a blend of oils, a semidrying oil, such as soybean oil, can be used to formulate materials in this group. These materials may be either knife or gun grade caulks. They will accommodate slight joint movement (about 10%) but have no recovery. They are slightly higher in price than the Class I Putties and are suitable for exterior glazing with wood or metal sash if painted regularly.

14.2.3   Class III—Skin-Forming Plastic Joint Sealants

This class includes most of the high quality soybean oil caulks as well as blends of other oils. Polybutene may be added to keep the mass tacky and flexible. However, the addition of tackifiers, such as polybutene, must be done with care: too much of this additive results in slow skin formation and a tendency of the caulk to pick up dirt. These sealants form a skin within a few days and then painters can follow. The bulk of the sealant bead then remains flexible or hardens very gradually over the years. These materials can tolerate movements up to 15 to 20%, but they have little or no recovery.

The filler most commonly used in the oil-based caulks is calcium carbonate. A typical high quality caulk might contain 60 to 75% calcium carbonate and 10% oil, with the remaining percentage made up in additives. Asbestos fibers added to the sealant increase its strength and help to resist the formation of drying cracks. Solvents are sometimes added to improve gunnability, but in general the oil-based materials are virtually 100% solids, resulting in very little shrinkage when the sealant is in place. Pigments are sometimes added, but the color range is quite limited. These caulks are available in natural, gray, aluminum, green, and redwood colors.

14.3 Properties

Since the oil-based sealants are designed for use in joints with little or no movement, there is very little published information about their properties. The oil-based caulks have no problems with odor or toxicity. They have poor resistance to many solvents and oils, but, if painted regularly, they can tolerate a great deal of intermittent water exposure. Their adhesive strength generally exceeds their cohesive strength so that, in the event of large joint movement, the material will rupture, rather than pull loose from the substrate. These caulks have fair to good UV resistance. Properties such as modulus of elasticity, ultimate elongation, and tear resistance are seldom measured since they are not required by the Federal specifications. Table 14.1 shows the advantages and disadvantages of the oil-based caulks as seen by the consumers.

14.4 Application

The oil-based caulks are furnished in both knife and gun grade and may be delivered to the job site in $\frac{1}{10}$-gal cartridges, or in 5-gal pails. For the homeowner market, these materials are available in cartridges and in pint and quart cans. The materials are easy to apply and require virtually no joint cleaning except for a quick dusting. Priming is desirable with wood substrates, but is unnecessary with other building materials.

Almost all users of caulking compounds use the oil-based caulks to

Table 14.1   Advantages and Disadvantages of Oil-Based Caulks
(as reported by consumers)

Advantages	Disadvantages
1. Good quality ones remain semi-plastic	1. Have little elongation and no recovery
2. Apply and tool easily	2. Poor durability—will dry out, become hard and brittle, crack, craze, and pull away from substrate
3. One-component systems	
4. No primers needed	
5. Lowest cost sealant or caulk	
6. Good wetting properties	3. Most experience a life expectancy of only 3–5 yrs
7. Relatively good color and color retention	
8. Higher quality ones have shown good durability (5–10 yrs)	4. Can withstand little or no movement (failure beyond a joint expansion of 5–10%)
9. Adequate performance where there is little movement as in housing and other low-rise buildings	5. Shrink and fail in adhesion with age
	6. Limited elasticity
10. Good package stability	7. Slow skinning and cure rate
11. Low initial shrinkage	8. Staining and bleeding of most substrates
	9. Must be painted regularly

some degree. The contractors use these materials for interior and exterior glazing with wood or metal sash, door and window frames, interior crack sealing, lap joint sealing in interior duct work, copings, and other applications. The homeowner uses these caulks for door and window frame caulking, glazing, gutters, downspouts, flashing, and general crack filling such as baseboards and plaster cracks. Some of these exterior uses by the homeowner are beyond the capacity of the oil-based caulk; this has led to some disappointments and has resulted in improved sales for better grade caulks such as the latex caulks and the butyls.

It is difficult for the consumer to know what he is buying when he attempts to purchase an oil-based caulk. Labels are difficult to pin down, but the consumer would probably find that the dealer carries an Economy Grade Caulk, an Architectural Grade Caulk, and perhaps a third known as an Interior Skinning Caulk. The Economy Caulk is the cheapest grade and is usually based on linseed oil (see Class I). The Architectural Caulk might be in either Class II or Class III, depending on ingredients, and would be somewhat higher in price. The Interior Caulk is a relatively new item containing a resin and a glossing oil so that it skins over very rapidly for almost immediate painting. This type of caulk would also probably be in the Class I price and performance range.

14.5 Summary

The oil-based caulks are nonelastomeric sealants designed for joints with little or no movement. They are virtually 100% solids, resulting in no shrinkage. These are the lowest price sealants on the market and are available in various quality ranges. They have been around for many years and have the advantage of familiarity. They are easy to apply, have no handling, storing, or mixing problems, and require no joint cleaning or priming.

Over the past 15 years, the oil-based caulks have lost a considerable segment of the market to the more exotic sealants. However, research has recently provided better grade caulks. Although the percentage of the market now occupied by the oil-based caulks will continue to decline, the overall total sales of these caulks will grow at a modest pace for many years. There are simply too many types of joints and cracks to be sealed by both contractor and homeowner in which no other sealant can compete on a price per performance basis.

15

Specialty Sealants

15.1 Introduction

Several other sealants have captured minor portions of the market. Some of these materials have developed because they have one or two special characteristics which are used to advantage. Others of these materials are as yet in the developmental stage. Newer and improved polymers being developed, together with improvements and crossbreeds of materials already in use, present a constantly shifting picture.

15.2 Neoprene * –Hypalon *

Polychoroprene (neoprene) sealants were among the first elastomers offered to the construction industry. The neoprene sealants can be formulated as both one- and two-component sealants. The one-component, gun grade sealants have a Shore hardness of about 35 to 40 and good cohesive strength. They will accommodate a moderate amount of joint movement (about 25%) with good recovery. The neoprene sealants also have very good water immersion resistance.

The disadvantages of the neoprene sealants are slow cure, application problems, and color. The one-component sealants, as presently compounded, are relatively high in solvent content (about 30%) and thus there is a great deal of shrinkage after the sealant is placed in the joint. The cure of the neoprenes is relatively slow. The material in some formulations can be difficult to extrude in cold weather. These are com-

* Trademark of E.I. DuPont de Nemours, Incorporated.

pounding problems which could be overcome if a suitable market were available. The major drawback of the neoprene sealants is the dark color. This factor alone rules out the neoprenes for most building sealing work.

The neoprene sealants do have certain properties which suit them for specialized uses. In highway construction, considerable research effort has been expended to develop the neoprene sealants as the lubricant-adhesive for use with preformed neoprene highway joint seals. Another market opening exists. The neoprenes have fairly good adhesion to asphaltic concrete, which is a rather unusual property. At the present time there are literally thousands of miles of portland cement concrete highways with asphaltic shoulders. Many knowledgeable highway engineers feel that this longitudinal joint between the pavement and the shoulder should be sealed to prevent the intrusion of water into the pavement subbase. The neoprenes, because of their adhesion to both portland cement concrete and asphaltic concrete, appear to be the logical candidates for this application.

In the consumer market, neoprenes are used for the repair of blacktop driveways, gutters, downspouts, and swimming pools.

The neoprene–hypalon sealants are relatively new to the market, having been introduced in 1964. The hypalons are somewhat similar in structure to the neoprenes, but they have excellent colorability and color retention. They also have excellent water immersion resistance. Because of the superior colorability, hypalon coatings are sometimes bonded to preformed neoprene seals for glazing joints.

The hypalon sealants cure much too slowly for practical use. A normal size building joint may take as long as three or four months to cure throughout.

The neoprene–hypalon sealants represent an improvement over each polymer, when used alone. The excellent water immersion resistance of these materials makes them one of the best available for the sealing of joints in swimming pools and pool decks. However, the slow cure, the poor extrudability, and the color have limited these materials to relatively few applications. Table 15.1 lists the properties of these materials as reported by the consumers.

15.3 Ethylene–Propylene

The ethylene-propylenes are generally formulated as a terpolymer, which leads to the nickname, EPT. These materials are at the present time mostly developmental, but they could become a major factor in the

Table 15.1   Advantages   and   Disadvantages   of   Neoprene-Hypalon
Sealants
(as reported by consumers)

Advantages	Disadvantages
1. Impervious to water 2. Remain flexible 3. Good elastomeric properties (elongation and recovery) 4. Good chemical, oil, ozone, and heat resistance 5. Good abrasion resistance	1. Extended cure time to reach elastomeric properties 2. Poor package stability (shelf life) 3. High degree of shrinkage (gun grade)

future. EPT can be vulcanized and preformed into a gasket shape. It can also be formulated as a gun grade sealant. The two chief advantages of EPT are the excellent weathering characteristics and the low price.

In the preformed sealant field, the limiting characteristic of EPT is its resistance to stress relaxation, or compression set. The preformed seal must continue to exert a high pressure against the joint walls in order to function. If the seal shows any appreciable amount of compression set, the seal has *failed*. Currently, neoprene, because of its excellent recovery, is virtually unchallenged as a preformed joint seal. However, it now costs $1 to $3 per linear foot to seal a highway joint with neoprene. If EPT could be formulated with sufficient recovery, it would capture a substantial share of this million dollar market.

As a gun grade sealant, EPT is currently the subject of research. The chief problems are reported to be a high viscosity resulting in poor extrusion from the caulking gun, and an adhesion demanding improvement. However, considering the number of other elastomeric compounds available, it seems certain that sealants of EPT in combination with other materials should be available shortly.

15.4   SBR

Styrene-butadiene rubber is another elastomer which has found only limited use. The chief advantage of SBR is its low cost. It offers moderately good elastic sealant properties at a very reasonable cost. However, its high solvent content (high shrinkage) and poor weathering characteristics have limited it to only minor usage.

15.5 *Polybutene and Polyisobutylene*

These sealants have three types of applications: they are used as gun grade sealants; as nonskinning, nonelastomeric tape sealants; and as tackifiers for other sealants.

As gun grade sealants, these polymers are noncuring. They remain permanently tacky. This can be quite an advantage when these compounds are used as bedding materials or in other concealed locations. Another advantage to the permanent tack of these materials is that they tend to self-heal if forced back together after a rupture. However, these sealants cannot be used in exposed locations because they pick up dirt and appear discolored, even though the color retention of the sealant itself is considered quite good. They are also easily picked or gouged from the joint by prying fingers. In overall performance, these materials are somewhat comparable to the butyls and the better grade latex caulks and have been largely supplanted by these two competitors. Actually, in terms of chemical structure the isobutylenes are quite similar to the butyls, since butyl rubber is a copolymer of isobutylene and isoprene. The properties of the polybutene sealants are shown in Table 15.2.

Table 15.2 Advantages and Disadvantages of Polybutene and Polyisobutylene Sealants
(as reported by consumers)

Advantages	Disadvantages
1. Good adhesion to many building structures 2. Maintain elasticity 3. Self-healing	1. Noncuring; remain soft and tacky 2. Nonelastomeric (poor elongation and recovery) 3. Poor solvent resistance 4. Low cohesive strength 5. Pick up dirt and dust 6. Stain porous substrates

As tape sealants, polybutene and polyisobutylene are in the class described by NAAMM Specification SS-lc-68 as "Nonskinning, Nonresilient Preformed compounds." These tapes are excellent glazing sealants (Fig. 15.1), but, because they are permanently tacky, they are often "topped off" with another elastomeric sealant.

In terms of total volume consumed, the largest use of these materials is as tackifiers for other sealants. These polymers combine well with

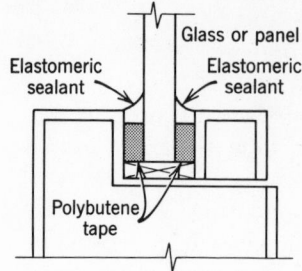

Fig. 15.1 A "topped-off" tape installation.

both oil-based caulking compounds and with other elastomers. The advantages they provide are good adhesion and, as compared with the oil-based caulks, improved elasticity and flexibility.

15.6   Rigid Sealants

The loose definition of a sealant allows any material placed in a joint to exclude water, air, and so forth to be called a sealant. In this context, the mortar used between bricks is a sealant. Although the mortar is rigid rather than flexible, it has good adhesion to brick and stone and good compressive strength. Therefore, for this particular use it is an excellent sealant.

There are other rigid materials which are used as sealants in construction, among which the most widely used are the epoxies and the polyesters. The epoxies have exceptional adhesion and are often used for the sealing of cracks in structural concrete. Figure 15.2 shows a structural concrete repair using epoxies. Both the epoxies and the polyesters are used in the sealing of pipe threads. The main differences between these materials are in water sensitivity and cost. The epoxies are not, in general, water sensitive. As a matter of fact, experimental composite beams have been formed by placing fresh concrete on the top of a steel beam which has been coated with epoxy. The epoxy and the concrete cure together and produce a bond.

In terms of cost, the polyesters are the lower priced materials. However, many polyesters in the uncured state are deteriorated by water. For dry applications, in which they can cure unimpeded by moisture, the polyesters can furnish almost as much adhesive strength as the epoxies.

Both of these materials have been used as coating sealants for concrete bridge decks. This application is discussed in greater depth in Chapter 27.

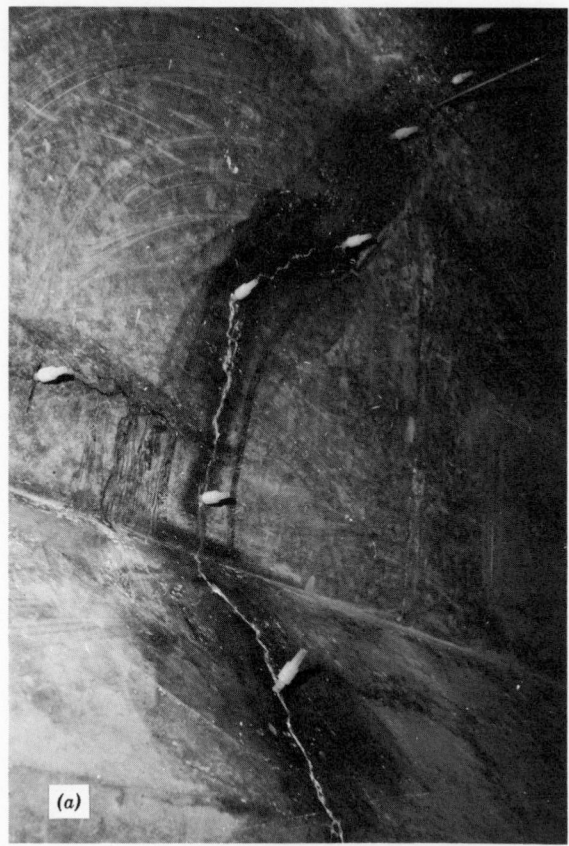

Fig. 15.2 Epoxy repair. (*a*) One-way valves placed into holes drilled along crack line at 10-in. intervals. (*b*) Quick setting epoxy gel seals crack surface and secures one-way valves. (*c*) Crack surface sealed, valves set, ready to grout. (*d*) Workman pumps high strength, low viscosity epoxy into lowest valve. After epoxy oozes from next highest valve, workman will move gun and repeat process until epoxy oozes from last valve. This assures complete penetration. (Courtesy of Sika Chemical Corporation)

The epoxies can be flexibilized or modified, with either coal tar or elastomers such as the polysulfides, to form special purpose sealants. These modified epoxies have excellent adhesion and can tolerate moderate amounts of joint movement (10 to 15%).

15.7 Roofing Sealants

It is questionable whether the roofing sealants should be classified as sealants or adhesives. They are generally bitumen-based materials,

Fig. 15.2 (b) Epoxy repair

highly filled with asbestos fibers. The hot poured materials are probably better classed as adhesives, but the cold applied solvent release materials are better classified as sealants. These cold applied sealants do not cure, in the strict sense of the word. Instead, they skin over rather rapidly as a result of the solvent release at the surface. The skin inhibits the release of the solvent in the remainder of the sealant mass. Consequently, the body of the sealant, under the skin, may remain tacky for several months.

The cold applied roofing materials are used for the sealing and waterproofing of cant strips, reglets, flashing, drains, and vent stacks during the roofing operation. The roofing sealants are not used in applications where appearance is a factor. As an example, in the installation of a cant strip for a built-up roof, the sealant may be used as a bedding material. After the cant strip is placed, more sealant is troweled over the top and bottom edges for waterproofing. No shape factor is considered

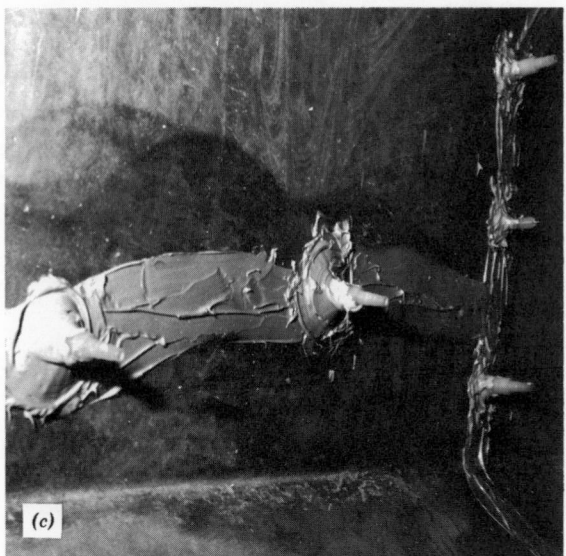

Fig. 15.2 (*c*) Epoxy repair

in this application because of the small movement involved. Conse-quently, the sealant is simply troweled into, and over the top of, the joints at the edges of the strip.

The cold applied roofing sealants are usually black and have a tack-free time of 4 to 6 hr. A well formulated roofing sealant will skin over, but will remain somewhat flexible under the skin. These materials have a tendency to soften with heat and become harder when exposed to cold weather. However, a good quality roofing sealant will not become brit-

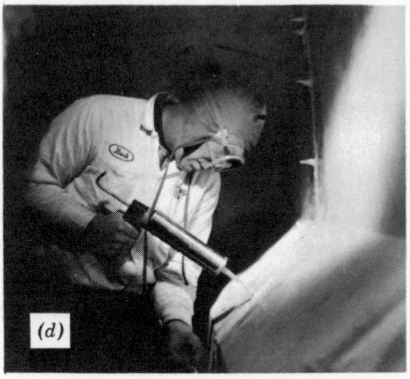

Fig. 15.2 (*d*) Epoxy repair

tle, even in very cold climates. These sealants are not elastomeric. They have virtually no recovery, but can tolerate modest movements (up to 10% of the adjoining parts. These materials have good adhesion. When failure occurs, it is generally of the cohesive type and is caused by aging and weathering embrittlement. Since the sealing of drains, stacks, cant strips, and so forth is an integral part of the roofing operation, this sealing is included in the roofer's bond for the completed job. Consequently, a service life of 15 to 20 years can be expected of a well compounded roofing sealant.

Other materials can do the job as well as the bitumen-based sealants, but in terms of price there appears to be little challenge. The roofing sealants sell in the range of $1 to $4 per gal and their modest share of the market appears relatively secure.

15.8   Sealants as Coatings

Up to this point, we have considered sealants to be materials placed into joints or cracks to preserve the integrity of a structure. However, certain construction applications for coatings may also be considered as sealants.

After several years of research, published reports [13] by the Highway Research Board show that linseed oil is the most effective seal coating for extending the life of portland cement concrete bridge decks and sidewalks. The linseed oil penetrates the pores of the concrete and helps to resist the deterioration by de-icing solutions and the freeze-thaw cycle. There is some slight discoloration and a tendency to dirt pick-up as the linseed oil dries, but this generally does not affect the overall appearance of the structure. Spray coatings of silicone are also useful for this application, and have worked well. However, on a price per performance basis, the silicones cannot compete with linseed oil in bridge construction.

In building construction, where appearance may be more critical, the additional cost of a spray coating of silicone may be justified to preserve the architectural beauty of a building. Elastomeric seal coatings have been applied directly to concrete shell and dome roofs, but the success of these coatings has been variable. The failures that have been reported are due not to poor coatings, but rather to an impossible stress situation. If a crack develops in the concrete, the coating which is bonded to the concrete must rupture.

Seal coatings are also used for driveways and bituminous concrete roadways. These coatings, sometimes known as slurry seals, are fairly effective and may be the only method of maintenance for areas which show extensive "map cracking." These seal coatings may be asphalt emulsions or solvent release bituminous materials. The slurry seals do a fairly good job of sealing cracks and enhancing the appearance of deteriorated asphaltic concrete. Their wear resistance, however, is not outstanding; therefore, in high density traffic areas, resurfacing the asphaltic concrete is a better alternative.

Other construction applications of seal coatings include the use of epoxies and silicones for encapsulating electric switches, connections, and bus bars. Other sealants are used as protective coatings for duct work and piping, especially at the pipe joints.

16

Preformed Sealing Tapes

16.1 Introduction

The preformed sealing tapes are actually extruded ribbons of sealant which are placed by hand. These preformed sealing tapes should not be confused with pressure sensitive adhesive tapes, such as the ordinary cellophane tape. The pressure sensitive adhesive tapes are discussed later in Chapter 26.

The preformed tapes were introduced to the construction industry in the early 1950's. The well known "rope caulk," which is familiar to most homeowners, began to appear in local hardware stores at about the same time. The tapes were slow in gaining initial acceptance, but their use has risen sharply within the past five years, and the projection is that they will grow approximately five times as fast as the overall sealant market within the next 10 years.

Preformed tapes act as both sealants and resilient fillers. Generally permanently tacky, they have become popular with contractors because they both eliminate the need for complicated installation equipment and also eliminate waste.

16.2 Compounding

The preformed tapes are produced in two general classes: the resilient and the nonresilient. The four basic polymers involved in tape formulation are polybutene, polyisobutylene, and cured or uncured butyl rubber.

The nonresilient tapes are compounded from polybutene or uncured butyl. Combinations of cured butyl, highly extended with polybutene, can also be used to produce a nonresilient tape.

The resilient tapes are compounded from either cured butyl or high molecular weight polyisobutylene. The resilient tapes are also combined with a tackifying agent (which may be polybutene) to provide the surface tackiness necessary for easy installation.

Many of the fillers used with the mastic sealants are also incorporated into the tapes. The type and amount of fillers vary with the intended installation. The ordinary rope caulk, which is a nonresilient tape, is highly filled with clay and a small amount of ground limestone (calcium carbonate). Reinforcing fillers, such as carbon black, are used in the resilient tapes. Asbestos, because of its fiber structure, is added to help the extrusions hold their shape until installation.

Pigment fillers are sometimes added to provide special colors for the tapes. Gray is by far the most common color for the tapes, although black, tan, green, and red are also available.

16.3   Properties

The preformed tapes are essentially a preformed strip of noncuring mastic sealant. Because these are noncuring sealants, they contain very little, if any, solvent and consequently have virtually no shrinkage after application.

The Shore hardness of the preformed tapes varies over a wide range. The hardness of the nonresilient tapes is so low that it is meaningless. The resilient tapes, on the other hand, are quite rubbery and may have hardness values as high as 30 to 40. The nonresilient tapes show little change in hardness with temperature changes. The resilient tapes may harden by as much as 10 to 15 points when the temperature drops below zero F.

The other kinematic properties of the nonresilient tapes are seldom reported. These sealants are actually more properly classed as joint fillers. They have enough surface tack to provide a seal, but very little actual adhesion. Consequently, properties such as adhesive strength and modulus have no significance. The nonresilient tapes also have almost zero recovery after being deformed. The National Association of Architectural Metal Manufacturers has a specification SS-1c-68 for the nonresilient tapes. Although brief, this specification is well written and adequate to measure the properties of these materials. It contains only four requirements: hardness, oil migration or staining, low temperature flexibility, and adhesion. Recognizing that the hardness values are too low for the Shore scale, NAAMM has specified the hardness in terms of a cone penetration test. The low temperature flexibility test consists of placing a bead of the sealant on a strip of mill finished aluminum. After conditioning for 4 hr at −30 F, the aluminum strip is bent 180° around a ½-in.-diameter mandrel. The sealant must not crack or lose adhesion during the test. The adhesion requirement of the specification requires that a bead of the sealant be placed on a steel plate and compressed under a 5-lb weight for 5 min. The steel plate is then inverted and a 2-in.-diameter steel ball is dropped on the plate from a height of 15 in. The sealant will show no more than 50% loss of adhesion as a result of this test.

The resilient tapes, because of their preformed shape, are generally tested by methods which are different from those used with the mastic sealants. The tapes are designed for specialized applications, and consequently are tested for properties which are relevant to the end use.

NAAMM Specification SS-1b-68 for the resilient tapes spells out requirements for hardness, adhesion, compression set, low temperature flexibility, and corrosion resistance, plus additional requirements for shelf life and accelerated aging and weathering. The adhesive requirements of this specification include both a tensile and shear adhesion test.

The adhesive and cohesive strengths of these materials are quite low and are roughly equal. A tensile adhesion specimen will generally fail cohesively at about 20 to 25 psi when tested at ⅛ in./min. Some of the resilient tapes are formed with reinforcing cords or fabric membranes. This reinforcement is placed longitudinally in the tape and prevents excessive stretch when the tape is placed in the joint. However, in an adhesion test the reinforcement is perpendicular to the direction of the applied load, which almost invariably insures a cohesive failure in the specimen.

The resilient tapes may be considered as moderate recovery materials. The NAAMM specification requires at least 75% recovery after compression under 10 psi for 24 hr. There is no specification requirement for the measurement of stress relaxation or recovery in either shear or tension. Both of these tests are conducted by manufacturers as "in-plant" evaluations, but no published data are available.

The low temperature flexibility test for the resilient tapes is quite similar to the test previously described for the nonresilient tapes. Cold specimens are bent 180° around a ½-in. mandrel to determine flexibility.

The preformed tapes are noncuring materials and they usually weather quite well. The NAAMM specification requires 500 hr in a weatherometer, but most of the high quality tapes go almost twice this long without evidence of distress.

There are many other types of tape seals which have found some limited application in construction. Cork neoprene tapes have been used for glazing. Asphalt-based tapes are manufactured, but, because they stain porous substrates and have little resilience, the market has been limited. There is currently some interest in the use of these low cost asphalt tapes for the sealing of sewer pipe. If this application proves successful, it could cut into the growth of the urethane pipe gasket. Tapes based on SBR and other rubbers have been produced, but have not received much acceptance.

16.4   Application

It is in the area of job application that the preformed tape sealants really stand out. These sealants require no mixing, no special equipment, and no messy clean-up. The tapes are delivered to the job in neat rolls. The tape is supplied with a paper backing which permits easy storage in roll form and easy installation. Typically, the tapes are about 1/8-in. thick and are available in widths up to 1 in. or more (Fig. 16.1).

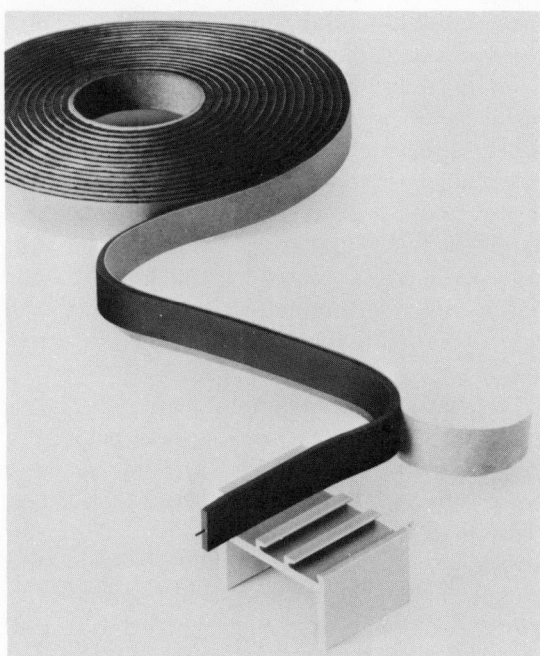

Fig. 16.1 A roll of preformed sealing tape. (Courtesy of Tremco Manufacturing Company)

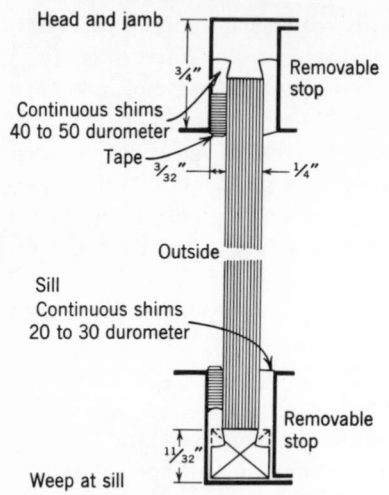

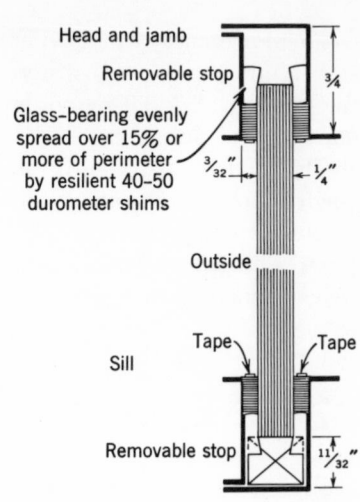

Fig. 16.2 Tape glazing installation. (Courtesy of PPG Industries, Inc.)

Fig. 16.3 Tape glazing installation. (Courtesy of PPG Industries, Inc.)

Keep in mind that the tapes are a more compliant form of preformed compression seal. The tapes have low adhesive and cohesive strengths and should not be used in working joints which have any significant degree of joint movement. They are also not suited for use in exterior sight-exposed joints, such as the one-stage weatherproofing of precast concrete or metal panel curtain walls.

The nonresilient tapes are used for hidden joints in wall construction. When used in glazing operations, they may be "topped off" with a bead of compatible gun grade sealant. These nonresilient tapes are also used in the installation of residential storm windows and storm doors. The homeowner uses these tapes for sealing cracks in masonry walls, bedding base molding and shoe molding strips to prevent drafty floors, sealing around window air-conditioners, and caulking exterior doors and window frames.

The resilient tapes find their major application in glazing. Figure 16.2 shows an application in which the glazing tape has been placed on the exterior of the window light only. This type of glazing operation employs a removable interior stop. Note that the resilient tape actually is a compression seal in flat, strip form. The interior stop forces the glass light against the tape, thus maintaining a tight seal. An important feature of this detail is that the entire glazing operation can be accomplished from inside the building. Since no exterior scaffolding is required, glazing from inside the building can result in considerable savings in the cost of the glazing contract. Figure 16.3 shows a detail in which the tape is used for both the interior and exterior seal. The removable stop is placed on the exterior side of the sash. Note, in both of these details, that the only purpose of sealing tape is to seal. The tape is not used to space the glass light in the rabbet. Shims and support blocks, as shown, keep the glass in place. Note also that there is an expansion space at the window head to accommodate thermal movement of the glass pane.

The preformed tapes also find widespread use in the lap joints of sheet siding and roofing. The tape in this application is held in position by metal screws or bolts through the adjoining parts, but under the action of thermal expansion and contraction the tape is subjected to a shearing type of movement. (See Fig. 1.1.)

The tape should be applied to clean joint surfaces. The tapes have a tacky surface and do not require the meticulous care demanded by the high elastomers, such as the silicones, urethanes, and polysulfides. However, the tapes should not be abused. They cannot be used successfully in wet or dirty joints.

A typical tape glazing job might be done as follows. First, the joint surfaces are cleaned. This cleaning is usually fairly simple because the sash, the glass pane, and the stop are separate and easily accessible pieces. Generally a solvent wipe, using a clean cloth and a nonoily solvent, is sufficient. Then, the correct lengths of tape for head, jambs, and sill are cut from the roll with scissors or a knife. The tape is placed against the sash with sufficient finger pressure to keep it in place. The paper liner is then peeled from the tape. Corner splices are very simply accomplished with enough finger pressure to push the tape ends together. Next, the glass light is set into place on the support blocks. Resilient shims are used to hold the glass in proper alignment. These may be U-shaped shims fitting over the edge of the glass or they may be small blocks of rubber with adhesive on one face. After the glass pane is properly aligned, the interior tape is placed on the perimeter of the glass. The tape liner is peeled off and the stop is snapped or screwed into place, and the installation is complete.

16.5 Summary

Because of their easy installation and low cost, the sealing tapes have experienced very rapid growth in the past five years, and should continue to grow even more rapidly. This growth has come mainly in glazing applications in the big construction field. The sealing tapes can be installed for approximately 15 cents per linear foot of joint. In the light construction area, such as residential work, motels, gas stations, small apartment units, and light industrial construction, the growth has been more modest. However, the trend in light construction is toward the use of more glass area and more sheet siding; consequently, this field of construction should be using tapes more frequently in the near future. The preformed tapes have found virtually no use in heavy, nonbuilding construction such as highways, bridges, dams, and airports. Homeowners are finding more uses for the sealing tapes, but most of these applications are remedial and small-scale; hence this usage will remain a growing, but minor, percentage of tape production.

17

Preformed Gasket Seals

17.1 Introduction

The concept of gasket sealing is as old as the problem of sealing. Any time man has had to mate two adjoining rigid parts in a leakproof fashion, a gasket is usually the first thought. In this very broad sense, every seal can be considered a gasket. Cork and metal gaskets are used in automotive work whereas in building work, mastic type sealants are actually gaskets that are formed on the job.

This discussion will be confined only to a certain type of gasket, namely, the preformed extrusions of high recovery rubber which are used for construction sealing. These preformed seals are sometimes known as compression seals because they are placed into a joint under compression and rely on interface pressure to maintain a tight seal.

As early as 1931 gasket sealing was attempted in highway construction. The seal used was a simple rubber tube, similar to a length of rubber hose. At that time, the seal was being placed along with a joint filler at the end of a slab unit. Considerable difficulty was encountered in keeping the seal in proper alignment, and therefore this attempt at gasket sealing was abandoned.

Preformed seals have been used in building construction for over 25 years. The prime application of the preformed seals has been in glazing work. One of the early developments in gasket sealing for buildings was an inflatable gasket made of synthetic rubber. This gasket fitted in a recess in the metal window frame. The sash was pivoted at the top and bottom to be opened for cleaning. When the gasket was inflated, the window was held tightly closed; when the gasket was deflated, the window could be swung open for easy cleaning from inside the building. This type of gasket was successful but expensive. Larger glass areas and less expensive extrusions forced it out of competition.

In the mid-1950's the preformed compression seals were again introduced into highway and bridge construction. By the following decade the compression seals had become the fastest growing seal for highway and bridge construction. Currently, the compression seals are probably used in more contracts for new construction than any other type of highway sealant. In the overall highway sealing market, that is, new work plus resealing, the compression seals rank second behind the hot poured sealants.

The preformed seals in building construction have shown a growth which is steady, but certainly not in proportion to the research and promotion money devoted to their development. The preformed seals certainly have many advantages, but, because of the physical disadvantages and high cost, many architects are reluctant to specify them. The preformed seals are used mainly on large building jobs. They are seldom feasible for light construction or residential work.

17.2   Compounding

The preformed compression seals have been formulated from neoprene, hypalon, butyl, silicone, SBR rubber, vinyl chloride polymers, and other materials. All of these are thermosetting materials, with the exception of the vinyls which are thermoplastic. Practically speaking, this means that the vinyls can be softened by the addition of heat and welded on the job; however, the others cannot.

Since the preformed seals must operate under compression, they must be formulated from high recovery materials. Because of its excellent recovery, neoprene has outstripped its competitors in both the highway and building construction markets. However, colorability is still a problem with neoprene, so that most of the neoprene seals produced are black. When colored gaskets are a necessity, neoprene with a bonded coating of hypalon may be used or other polymers may be specified.

Most of the preformed seals contain at least 50% polymer. Carbon black is the most widely used filler, both for its reinforcing value and because it helps the UV resistance of the seal. With polymers other than neoprene, other fillers such as calcium carbonate are sometimes used, especially in the compounding of colored seals.

17.3 Properties

The properties of the preformed gaskets for building construction as observed by consumers are shown in Table 17.1.

Table 17.1 Advantages and Disadvantages of Preformed Gaskets
(as reported by consumers)

Advantages	Disadvantages
1. Require the least labor of all sealing methods 2. No waste 3. Significantly reduce or eliminate the need for sealants and caulks 4. Expected long life	1. Tolerances under which they are manufactured are insufficient to insure leakproof joints, especially in precast concrete (some architects and contractors are reluctant to specify and use the gaskets, respectively, because of the uneven contact between the gasket and the concrete) 2. Joints must always be in compression to insure sealing 3. Gaskets must be tailored to the different size windows around which they are used 4. Field welding of joints requires the use of a special adhesive 5. Low temperature rigidity 6. Color limitations (for exterior usage only black can be used because of the need for good aging and weathering characteristics) 7. Separate pieces must be joined in a water-tight bond and thus the corners of the gaskets can be troublesome 8. Relatively expensive (20–45 cents per linear foot versus 4–12 cents for premium sealants)

17.3.1 Neoprene

The neoprene seals are available as both dense extrusions and as cellular sponge extrusions. The neoprenes can be extruded to close tolerances in both the dense and sponge forms. The neoprenes have good resistance to oil, water, and a wide range of mild acids and solvents. The resistance of the neoprenes to UV radiation and ozone is excellent.

These seals are delivered to the job site as fully cured materials and thus they have no problems with odor or toxicity.

The Shore hardness of the dense extrusions ranges from 40 to 60. This value of hardness will increase by 10 to 20 points as the temperature drops below zero F. This stiffening of the neoprene makes cold weather installation quite difficult. Also, this high value of hardness prevents the seal from accommodating itself to irregularities in the joint wall. Consequently, the preformed seals are excellent glazing seals, but are of questionable value for the sealing of joints in concrete or stone wall panels.

The tensile strength and ultimate elongation of neoprene are very good. Precise values of these quantities are important quality control measures for the manufacturer, but are of little value to the consumer in terms of compression sealing. Hence, they are not reported here.

The outstanding property of neoprene, which accounts for its position of leadership in compression sealing, is its high recovery. The neoprene seals can be compressed for long periods of time with only slight evidence of stress relaxation. There are glazing installations of dense neoprene compression seals which have been in place for 20 years and are still functioning effectively.

The cost of the neoprene extrusions is high in relation to other sealing materials. In highway work, for instance, it costs 10 times as much to seal a joint with a compression seal as with hot poured asphalt. In building work, it costs approximately 45 cents per foot to seal with a compression seal, as compared to 15 cents per foot for a premium gun grade elastomer.

The neoprene sponge extrusions are closed-cell foams with a thin impervious skin. These cellular extrusions are quite soft and compliant and can be used successfully in joints with irregular joint walls. Since these seals are sponge materials, they have no measurable Shore hardness. The foams will stiffen somewhat as temperature drops, but they can be easily installed at almost any practical temperature. The resistance to water, oil, UV, and ozone is comparable to the dense neoprene extrusions.

The recovery of the cellular neoprenes is good but not outstanding. The cellular neoprenes should not be installed under more than 40% compression [7]. If installed under greater compression, the recovery drops off rapidly. Both laboratory and field results show the sponge neoprenes functioning after a two-year period, but these are still relatively new materials and do not yet have the long-term field history of the dense extrusions.

17.3.2   Vinyl

The vinyl extrusions are thermoplastic materials. In addition to the obvious advantage of field weldability, the thermoplastic nature of the material means that it can be easily extruded to complicated shapes with good dimensional control. However, the thermoplastics have very little chemical cross-linking which means they have a tendency to flow under stress, over the whole temperature range.

The vinyl extrusions are generally formed in a Shore hardness between 60 and 90. This value increases by 10 to 20 points as the temperature drops below zero F. At temperatures below −20 F, the vinyl seals become almost brittle.

These vinyl seals have good resistance to ozone, weathering, chemicals, and oils. They are produced in a range of colors to suit various sealant applications. The color retention of the vinyl extrusions is very good, even when subjected to extended weathering.

The vinyl seals have good tensile strength and elongation, and very high tear strength. However, they are low recovery materials and should not be used in working joints. These seals exhibit a great deal of stress relaxation. Consequently, they tend to relax and flow under sustained loads and thus do not form an effective compression seal for working joints.

The vinyl seals are very convenient to use. They are usually shipped to the job in 50-ft coils and require no special storage or handling. They are easily cut to exact lengths (whereas the more rubbery seals are not) and, because of their high tensile strength, are subject to very little stretching during installation. Field splicing of the vinyl seals can be accomplished with a hot soldering iron. Corner joints which are particularly troublesome for many of the other preformed seals can be accomplished with a miter cut and field welding.

The vinyl seals are roughly in the same price range as the neoprene compression seals; that is, 40 to 45 cents per foot of joint.

17.3.3   Butyl

The preformed butyl seals are extrusions of cured or vulcanized butyl. These are thermosetting materials and are a little more difficult to extrude to close tolerances than some of the other seals. The butyls have been produced in sponge form, but these cellular butyls have received little acceptance. Almost all of the preformed butyl seals are the dense extrusions.

The butyl seals are usually produced in the 30 to 40 Shore hardness range. This value of hardness will change by 10 to 15 points at cold temperatures. The butyls can be produced in a range of colors, but black and gray seals are by far the most common. The color retention of the butyls, after extended weathering, is only fair. The butyls have good resistance to weathering, ozone, and water, but only mediocre resistance to oil.

The butyl seals may be considered as moderate recovery materials. Their recovery is better than the vinyl seals but not nearly as good as the neoprenes. The butyl seals can be successfully used in working joints with small to moderate (up to 20%) movement, but should not be used in expansion joints or control joints which have movements in excess of 25%. The butyls, although somewhat softer than the dense neoprene extrusions, are still not compliant enough to waterproof joints with irregular joint walls, such as the joints between concrete or stone wall panels.

Regarding field handling properties, the butyls are similar to the neoprenes. They stiffen somewhat at cold temperatures and thus become harder to handle. They are more difficult to cut to exact length than the vinyls. Field bonding can only be done with special adhesives.

The cost of the butyl seals is somewhat lower than the neoprenes. The preformed butyl seals can be installed for approximately 30 to 35 cents per foot of joint.

17.3.4 Silicone

The silicone gaskets are high quality preformed seals. The silicones have very good colorability and color retention. They also have good aging and weathering properties and good resistance to oils and solvents. The recovery of the silicones is excellent. In summation, the silicones have all the properties required of a high quality compression seal; unfortunately, they also have a high price tag. Assuming labor costs are identical for both materials, it costs over twice as much to seal with silicone as with neoprene.

17.3.5 SBR

Styrene-butadiene rubber (SBR) compression seals have not been widely used. The chief reason for the lack of acceptance has been their inferior weathering characteristics. The recovery is good and the cost is reasonable, but these seals have poor color retention and poor resistance to UV and ozone. Consequently, they are not suited for use in exposed

Fig. 17.1 Preformed seal shapes. (Courtesy of D. S. Brown Company)

locations. They may, however, be used successfully where weathering is no problem.

17.4 Application

The applications of preformed compression seals in building construction can be subdivided into four general classifications.

> Glazing seals
> Seals for exterior panel joints
> Structural gaskets
> Miscellaneous preformed seals

Note that the seals in all the above classifications are compression seals which function by maintaining an interface pressure against the joint wall. Preformed seals for highway and bridge construction are not included because they are discussed in Chapter 19.

17.4.1 Glazing Seals

The glazing seals are manufactured in a multitude of shapes to fit specific applications. Some of these shapes are shown in Fig. 17.1. The most common shape of compression seal for glazing is the U-shaped channel gasket. These compression seals are usually extruded either from cellular neoprene or else in a relatively soft grade of dense extrusion.

The channel gaskets may be designed for either shop or field installation. Shop applications may be either the foam or the dense extrusion. In field applications, the cellular gaskets are favored. For shop applications, the dense extrusions may be cut to length, fitted with molded corners, and vulcanized into a continuous loop which can be fitted around the perimeter of the glass pane. For field applications, molded corners of cellular gasket may be fitted to the glass pane. The straight runs can then be cut to length. It is best to cut the straight runs about ¼ to ½ in. longer than the measured distance, to compress the sponge gasket slightly in a longitudinal direction during installation. Joining the straight runs to the corners is accomplished with an adhesive. When the stops are installed, the gasket is compressed about 25 to 30% in the lat-

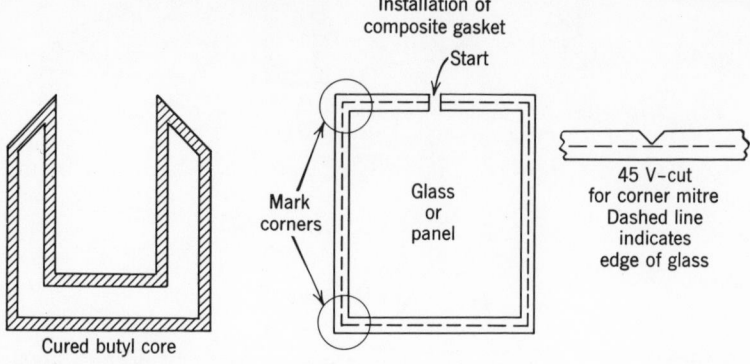

Fig. 17.2 Composite butyl gasket. (Courtesy of Enjay Chemical Company)

eral direction, and the installation is complete. Note that the neoprene sponge extrusion is not used to support the glass pane or to space the glass pane in the rabbet. Support blocks and shims are used for spacing and thus the sponge extrusion performs the sealing function only.

Extruded ribbons of cellular neoprene are also used for glazing. These strips are installed much like the sealing tapes. Straight sections are generally joined at the corners with a miter cut. These corners may be troublesome and, hence, are sometimes touched-up with a small amount of elastomeric sealant.

The composite gasket, which is formed of a core of cured butyl with an outer layer of soft, uncured butyl, is installed in a slightly different fashion (Fig. 17.2). The straight length of seal is cut first. Then 45° miter cuts are made for the corners and the gasket is fitted around the glass. This seal has disadvantages. It is difficult to provide proper support and spacing for the glass pane; also, the exposed portion of the seal is soft, uncured butyl which is subject to dirt pick-up and vandalism. Consequently, this composite gasket has received little acceptance.

17.4.2   Seals for Exterior Panel Joints

Compression seals for the one-stage weatherproofing of exterior wall panels are generally similar in cross section to those used in pavement joints, as shown in Fig. 19.16. These seals have not received wide acceptance because of installation problems. It is difficult to compress the seal and insert it into place properly in a vertical wall.

Sponge extrusions of rectangular cross section, considerably easier to install than the dense extrusions, are just now beginning to receive some attention.

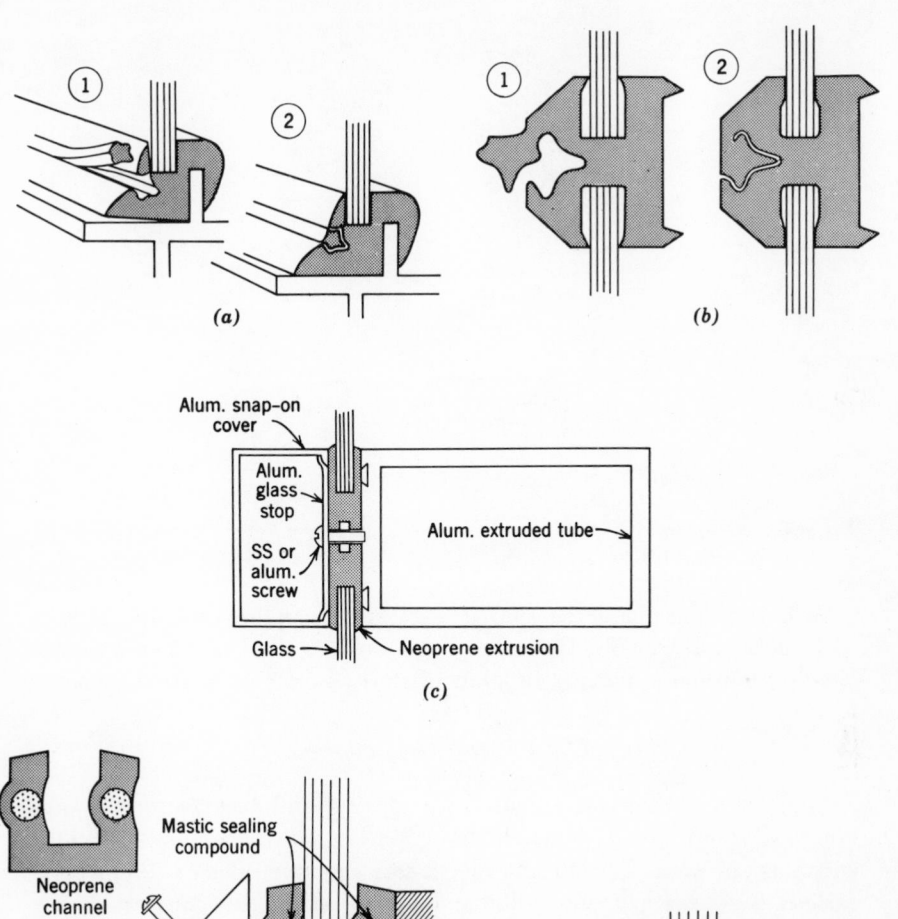

Fig. 17.3 Structural glazing gaskets. (a) Inlock strip. (b) General American Transportation Co. strip. (c) Olin Mathieson channel. (d) Pawling rubber wet seal. (e) Pawling rubber wet seal. (Courtesy of Pawling Rubber Company)

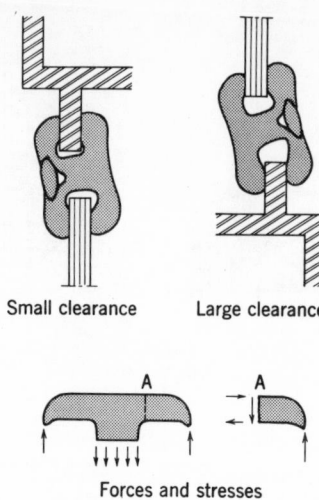

Small clearance   Large clearance

Forces and stresses

Fig. 17.4 Action of structural gasket under load. (top) Action under wind loads. (bottom) Internal stresses.

Both the dense and the sponge extrusions can be used successfully with metal wall panels. However, the sponge is the better performer in panel-to-mullion joints and in joints of stone, concrete, or brick masonry.

17.4.3   Structural Gaskets

The structural gaskets are used for glazing and they perform a dual function. These seals support the glass, in addition to providing a weathertight seal. The structural gaskets are sometimes called zipper gaskets because they quite often include a zip-in strip to provide the necessary compression. Figure 17.3 shows several patented types of structural gaskets. The structural gaskets are generally formed of harder, stronger neoprene than the other extrusions. Structural gaskets have a hardness of 70 to 75, as well as a high tensile strength. Figure 17.4, a cross section sketch of one such gasket as analyzed by Oberdick, illustrates why the extra strength is necessary. These gaskets usually have molded corners and are vulcanized into a continuous gasket for each pane of glass. As of this time, there is no structural adhesive for on-the-job use which will join these gaskets together into a strong, leak proof seal.

The structural gaskets are covered by ASTM Standard C542. This is a well written specification with requirements for hardness, tensile strength, and ultimate elongation which are all important properties in a structural gasket. The most important section of this specification is a requirement for the lip sealing pressure that the gasket must exert on the glass. The value of lip seal pressure required is 4 lb per linear inch. However, this pressure is measured only on new gaskets. The specification does not require a minimum value of lip sealing pressure after aging. Consequently, the quality of the neoprene and the design of the seal must be excellent if the seal is expected to function for 20 years.

17.4.4 Miscellaneous Preformed Seals

These seals come in a multitude of shapes for specialized applications. Some of the glazing seals are shown in Fig. 17.3. Other special shapes are used for pipe gaskets, continuous door stops, and wiper-type gaskets for movable sash. The structural gaskets have found some application in the sealing of heating and air-conditioning ducts.

Many of these special seal shapes are designed for plant application and may be part of the finished building component that is shipped to the job site. An example would be a large metal curtain wall panel, complete with fixed light glazing. One example of a field application of

Fig. 17.5 Cap seal. (Courtesy of D. S. Brown Company)

one of the specialized seal shapes is the cap seal shown in Fig. 17.5. This cap seal forms a neat joint in exposed interior wall panels.

17.5 Summary

The preformed compression seals offer a generally high quality, but high priced, solution to the sealing problem. The preformed seals have a very good life expectancy, in the neighborhood of 20 years. These gaskets also offer a very neat looking joint with a minimum of on-the-job labor. There is little or no waste with the preformed seals and a minimum of operator's clean-up time.

However, the preformed seals have some disadvantages which have limited their growth in construction. The seals must be accurately sized for the specific joint they are to seal. While the gaskets themselves can be extruded to fairly close tolerances, the fabrication and erection tolerances of the adjoined parts may make the seal ineffective. If the seals are shop-vulcanized into a continuous gasket, the cost is high. If the seals are joined in the field, the junction of the separate parts into a leakproof union remains a problem.

Curtain wall construction has grown quite rapidly and should continue. The overall trend in construction is toward more prefabrication, which means more use of preformed gaskets. The preformed seals over the last 10 years have shown a slower growth than the overall sealant market.

18

Waterstops

18.1   Introduction

A waterstop is a specialized type of preformed seal which is used for the waterproofing of concrete structures. The waterstop is actually a diaphragm which is cast as an integral part of the concrete and bridges the gap between concrete units. Waterstops are used in construction joints in walls and slabs; for example, the corner joint where a concrete wall meets a foundation, and many other uses.

Waterstops are widely used in the industrial or big construction market. They are also widely used in heavy nonbuilding construction, such as tunnels, dams, retaining walls, swimming pools, and culverts.

18.2   Materials

Waterstops are a particularly demanding application. In building construction, if a sealant fails and the building leaks, the building can be recaulked. However, the waterstop is placed as an integral part of the concrete construction. Waterstops cannot be replaced; they must last for the entire life of the structure. Consequently, the choice of a material for a waterstop is a critical decision.

Waterstops have been fabricated from a wide variety of materials, including neoprene, polyvinyl chloride (PVC), and other plastics and rubbers. Formed strips of copper also make a very good waterstop for many applications.

The material used for a waterstop depends on the particular application. The waterstop is generally selected on the basis of the type and amount of movement it must accommodate. Most waterstop applications are designed for low to moderate amounts of movement (up to 15%). Types of movement are about equally divided between tension and shear movements, depending on the particular design.

The vinyl extrusions have become by far the most widely used form of waterstops. These PVC waterstops, like the preformed seals, can be extruded to very close tolerances. Typically, the PVC waterstops have good tensile strength (1800 psi) and excellent tear strength (250 lb/in.). The material itself has an ultimate elongation of about 300%, but, when fabricated to shape and subjected to simulated service loads, it has an elongation capacity of about 20%. The vinyls are thermoplastics and, consequently, tend to stiffen at cold temperatures. Fortunately, the material does not become brittle until the temperature reaches approximately −30 F. Since the waterstops are embedded in and protected by a few inches of concrete, brittleness has not been a problem. In construction joints and other joints where minimal movement is encountered, the PVC waterstops are the most widely used.

In isolation joints or control joints, an elastomeric material is probably a better choice than PVC. Neoprene has been used quite successfully in the production of waterstops. The waterstops should be a natural application for ethylene-propylene (EPT) compounds. EPT has outstanding weathering characteristics and is low priced. The resistance of EPT to compression set is still somewhat doubtful, but this is of little or no consequence for waterstop applications. The neoprene waterstops have roughly the same tensile strength and ultimate elongation values as the vinyls, but the neoprenes are more rubbery and are better suited for applications in these working joints.

As with the preformed seals, the vinyls are field weldable, whereas the neoprenes are not. Copper strips can be joined on the job as easily as the vinyls.

18.3   Application

Waterstops are fabricated in hundreds of individual shapes, but with one central theme. The waterstop consists of two wing portions which

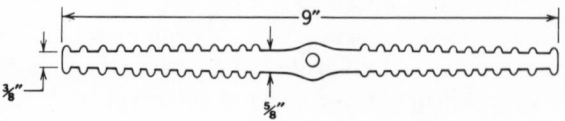

Fig. 18.1 Waterstop with central bulb.

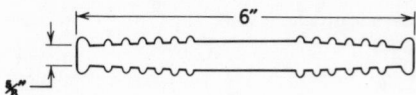

Fig. 18.2 Waterstop with no central bulb.

are embedded in the concrete and a central portion to accommodate the movement. The waterstop shown in Fig. 18.1 has a hollow bulb in the center and is designed to accommodate moderate movements. The serrations on the wings of the extrusion help bond the seal into the concrete and help the watertightness of the seal. Figure 18.2 shows a waterstop with no central bulb. This unit is designed for construction joints and other joints which encounter little or no movement. Both of these seals have been formed in neoprene and PVC. The copper waterstops are formed from long strips of copper sheet (Fig. 18.3). The wings of the waterstop are flat and rely on the bond between copper and concrete to obtain a waterproof seal. The movement between the adjacent concrete units is accommodated by the V-bend in the strip.

The waterstop may be completely responsible for the watertightness of a joint, or it may be used as a second line of defense in conjunction with another sealant.

The typical application of a waterstop involves embedding one wing portion of the seal in the concrete, leaving the other wing portion exposed. As construction proceeds, this exposed portion of the waterstop is encased in the adjacent concrete placement, and the installation is complete. Waterstops of copper or vinyl sheet can be joined in the field as shown in Fig. 18.4.

A few joint details, in which waterstops have successfully been used, are shown in the following figure. Figure 18.5 shows a joint between a base slab and a foundation wall below grade. In this case, the waterstop is cast into the base slab with one wing portion exposed. As the foundation wall concrete is placed, it encases the upper portion of the waterstop. In this application, care must be taken to insure positive alignment and contour of the waterstop and to insure that it is not bent over and rendered ineffective by the weight of the fresh concrete pouring in on

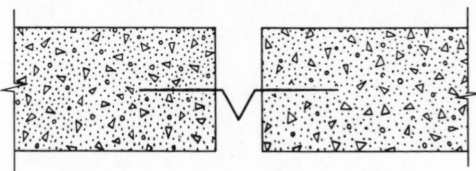

Fig. 18.3 Copper waterstop.

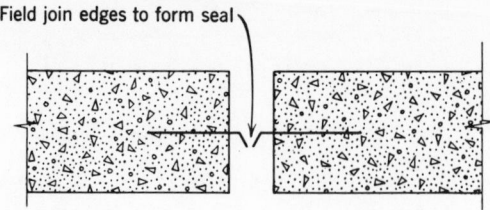

Fig. 18.4 Field joined copper or sheet vinyl waterstop.

top of it. Note that in this application any relative movement between the adjacent portions of concrete places the waterstop into shear.

Figure 18.6 shows the waterstop in the base slab of a swimming pool. In this application the waterstop is used together with another sealant. Relative movement between the adjacent slabs may be either tension

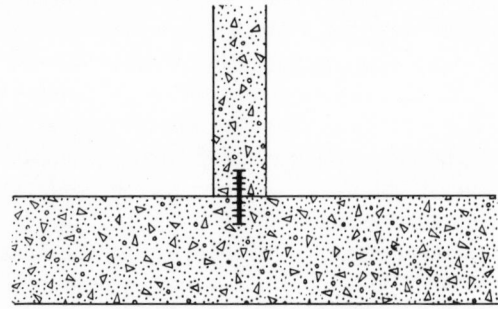

Fig. 18.5 Waterstop installation.

and compression due to thermal change, or shear caused by differential settlement.

Figure 18.7 shows a copper waterstop used with another sealant to waterproof the isolation joint in a retaining wall. Again, the movements may be either tension-compression or shear.

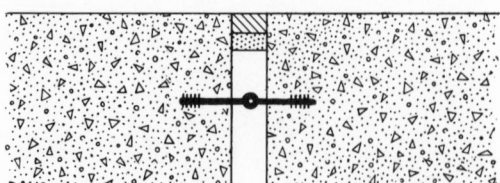

Fig. 18.6 Waterstop in the base slab of a swimming pool.

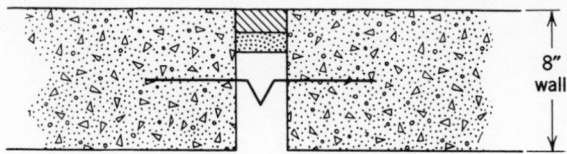

Fig. 18.7 Section through the vertical joint in a retaining wall. Copper waterstop used with sealant.

18.4 Summary

Waterstops can form an excellent barrier to prevent the movement of water through the many construction joints in concrete structures. Waterstops, once installed, cannot be replaced; consequently, material must be of the highest quality, and installation procedures must be very carefully supervised.

Waterstops are available in a wide variety of shapes. The materials most commonly used are PVC, neoprene, and sheet copper. In recent years, the use of PVC and neoprene has increased, whereas the use of copper has decreased proportionately. Although costs vary somewhat with materials, the installed waterstop should cost between 25 and 40 cents per foot of joint.

19

Sealants in Highway Construction

19.1 Introduction

The sealing of joints in highway pavements is one of the most trouble-some problems facing highway engineers today. The old concept of "pour some tar in the joint" is simply not acceptable on today's high speed expressways. The annoying "thump-thump" of automobile wheels over the pavement joints not only makes a bumpy ride, but also makes the vehicle harder to control at high speeds. Most motorists are aware that the annoying thumps shorten the life of automobile tires and sus-pension systems, but few of them realize that the thumping is costing taxpayers millions of dollars in highway and bridge maintenance and re-pairs.

Highway engineers are concerned with two major factors: highway safety and low cost per year over the life of the pavement structure. Properly sealed pavement joints not only will increase highway safety, but also will result in lower maintenance costs for the system. Another minor factor the highway engineers are beginning to consider is the sim-ple "cosmetics" of highway construction. A highway with sticky sealant extruded out of the joints and tracked over the pavement is a sloppy ap-pearing pavement. Today's taxpaying drivers are demanding not only a safer, smoother ride, but also better looking highways.

19.2 History

"The proper sealing of construction, contraction, and expansion joints is a problem that has defied practical solution since the beginning of

166

paved roads." [14] Early Roman roads were built of large aggregate and paving blocks which absorbed the movement of the pavement over a large number of joints. In this country, the growth of the cement industry spurred the development of rigid pavements; as early as 1912 in fact, work was being done on joints. The first rigid pavements, of course, were built without joints and thus the pavements developed a random pattern of cracking. Consequently, the joint was introduced to control the cracking. In the early days of joint sealing, trial and error was the rule and experience was the great teacher. Early joint sealing materials included soft wooden boards, tar paper, sand, and tar.

As early as 1930, the cement industry, as a group, began to take an active interest in joint sealing, and the first efforts were made to provide a scientific basis for pavement joint sealing. Both expansion and contraction joints were used, dowels were commonly used, and definite answers were being sought to the problems of joint sealing.

By the late 1930's and early 1940's the state highway departments had acquired a considerable degree of sophistication. The states began to undertake studies of pavement movements and to record the performance history of pavements.

In 1960 the American Association of State Highway Officials began the most intensive program of highway research ever attempted. The results of this famous AASHO Road Test were published in several separate volumes. A great deal of work on pavement movement was included in this research program.

After World War II, the construction boom forced the development of more research into pavement behavior and also aided the development of newer joint sealing materials. Rubber joint materials had been used as early as 1931, but had been largely unsuccessful in practice. Hot poured asphalts were the most widely used sealing materials in the 1940's. The early 1950's saw the introduction of the cold poured elastomers. The polysulfides led the parade, to be quickly followed by other elastomers.

In recent years the old rubber tube concept of 1931 has been resurrected and given a new look. Preformed gaskets of neoprene have been used with considerable success in pavement joints. Another recent development is the "presealed joint." This device consists of two stainless steel plates and a neoprene gasket which is precompressed to a given width. This device is vibrated into the fresh concrete as the final step in the paving process.

In order to stimulate and fund highway research, the American Association of State Highway Officials, the Bureau of Public Roads, and the Highway Research Board have formed the National Cooperative Highway Research Program. This program has included several large-scale projects on highway and bridge behavior and also projects on joint sealing materials.

To coordinate the publication explosion, NCHRP sponsors the Highway Research Information Service, which provides an accurate record of research in progress and also furnishes a computerized search of the literature for research in any aspect of the highway research field.

19.3   Why Seal the Joints?

Joints are expensive. They cost money to form or saw; they are difficult and expensive to seal, and more expensive to keep sealed. Why not simply build the pavement without joints and eliminate the problem? After all, the railroads use continuously welded rail a mile or more in length; so why can't we do the same thing with concrete pavements?

In order to construct a pavement without joints or cracks, the pavement must be heavily reinforced. A continuous pavement constructed with the normal amount of reinforcement simply cracks and forms its own joints. There is a great deal of research being conducted into continuously reinforced pavements. Many test sections of this pavement have been built. Some of these heavily reinforced pavements have been very successful, but others have not fared as well. The prime drawback to the use of continuously reinforced pavement is cost. These pavements demand a large quantity of steel and the highest quality of construction. The use of continuously reinforced pavements will undoubtedly grow, but the cost will probably preclude its use in any but the highest type of pavements. Prestressed concrete pavements have also been installed on an experimental basis in several states. However, cost will probably limit their use to the same applications as the continuously reinforced pavements. Consequently, jointed pavements will probably be with us for many years to come.

Joints in highway pavements are sealed to prevent the intrusion of incompressible solids, water, and highway deicing chemicals. All types of slab openings are sealed: contraction joints, expansion joints, longitudinal joints, and random cracks. Water and salt solutions seeping into the joints corrode the load transfer devices between slabs, which inhibits normal pavement movement. Water seeping into pavement cracks corrodes the reinforcement, which reduces the strength of the slab. Water passing through the joints also may result in the softening of the subgrade under the slab edges. This loss of subgrade support magnifies the

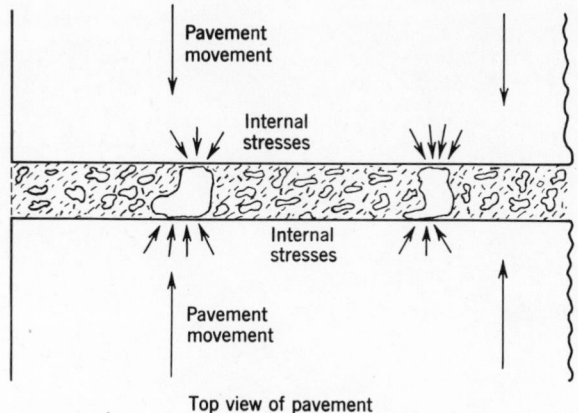

Top view of pavement

Fig. 19.1 Slab stresses due to incompressibles in the joint.

pumping effect of traffic passing over the joints and results in the cracking of slab edges and corners.

Solid material working its way into pavement joints is even more serious than water. In winter the slabs contract and the joints open. If not
sealed, the joints fill with stones and road dirt. With the coming of
warmer weather the slabs try to expand, but they have no place to go.
Because the foreign matter filling the joint is not uniform in size, con-
centrated stresses are built up in the slab edges as the slab attempts to
expand (Fig. 19.1). These stress concentrations result in spalling of the
slab edges, which accelerates the deterioration caused by water and deicing chemicals. This spalling is unsightly and makes the joint more difficult to seal in the future. Spalling is not confined to the tops of slabs.
Many pavements are built on a granular subbase; as the pavement expands
and contracts, granular material is scooped into the joints and causes spalling. Studies in several states have shown that spalling at the bottoms of
pavement joints is at least as serious as the visible surface spalling.

Another problem associated with incompressible material in the joints
is pavement growth. Pavement growth is undoubtedly the composite effect of a large number of contributing factors, but the joints are felt to
be a large part of the problem. Incompressible material in pavement
joints can prevent normal pavement movement. A pavement which is
prevented from normal movement may actually explode upward (Fig.
19.2) or the whole pavement mass may move longitudinally. This pavement translation is not an isolated occurrence. A recent study [15] by a
research group at the University of Mississippi showed that the problem

Fig. 19.2 Pavement blowup.

Fig. 19.3 Bridge seat tilted and cracked by pavement growth. (Courtesy of Acme Highway Products Company)

Fig. 19.4 Joint filler extruded upward by pavement growth. (Courtesy of Acme Highway Products Company)

exists, at least to some extent, in 80% of the continental states. Movements caused by this pavement growth may amount to as much as 1 or 2 ft. As an example, consider a pavement with joints spaced at 60 ft. This spacing gives 88 joints per mile of pavement. If just ⅛ in. of incompressible material finds its way into these 88 joints, the total movement that must be accommodated is 11 in.

Pavement translation may result in the displacement and misalignment of curves, or it may be accommodated at the nearest bridge structure. The moving slab often tilts and cracks the backwalls of bridges, pushes bridges out of skew, or in extreme cases actually pushes the bridge off the bridge seat. Figures 19.3, 19.4, and 19.5 show some of the damage caused by pavement growth.

19.4   Types of Joints

Joints in pavement construction fall into two general categories: contraction joints and expansion joints. The contraction joint is a weakened plane in the slab, which is produced either by forming or sawing. The depth of the cut is usually ⅙ of the slab depth. The expansion joint is a full depth joint through the slab. Construction joints, which occur at the end of a day's paving operations, are full-depth joints through the slab. A day's construction is generally terminated at the end of a slab unit, so that the construction joint is, in effect, another expansion joint or contraction joint, depending on state practice.

Fig. 19.5 Skew bridge pushed 2 in. out of alignment by pavement growth. (Courtesy of Acme Highway Products Company)

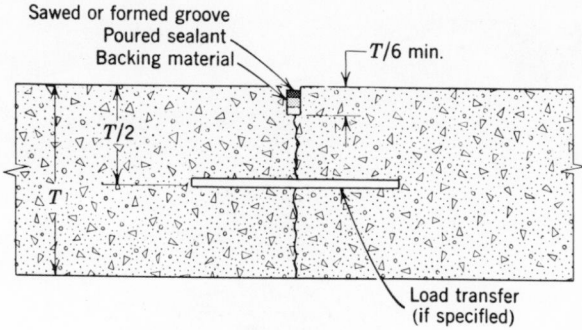

Sawed or formed groove
Poured sealant
Backing material

$T/6$ min.

$T/2$

T

Load transfer
(if specifled)

Fig. 19.6 Contraction joint.

19.4.1 Contraction Joints

The contraction joint (Fig. 19.6) is a transverse joint which is sawed or formed to create a weakened plane in the slab. The opening is generally $\frac{1}{6}$ of the slab depth. As the drying shrinkage takes place in the concrete, the slab cracks throughout the remainder of its depth. After the cracking has occurred there is still some aggregate interlock, but usually a load transfer device is placed in the slab to prevent relative vertical movement between slab edges. Several patented types of load transfer devices are available (Fig. 19.7), but the straight dowel bar (Fig. 19.8) is still the most common type of load transfer device in use. The contrac-

Fig. 19.7 A patented load transfer device. (Courtesy of Acme Highway Products Company)

Fig. 19.8 Dowel bars in place at a pavement joint. (Courtesy of Acme Highway Products Company)

tion joint is generally ¼ to ½ in. wide: the ⅜- in. joint is the most common. The width of the joint should depend on the joint spacing: the longer the slab unit, the wider the joint opening should be. Joint spacings also vary quite widely from state to state. New York State uses a spacing of 60 ft 10 in., while California uses a staggered spacing of 13, 19, 18, 12 ft. The staggered spacing is effective in reducing any rhythmic thump under the wheels of the vehicle. Several states are also experimenting with skewed contraction joints to distribute wear and reduce wheel thump.

Pavements nowadays are seldom constructed one lane at a time.

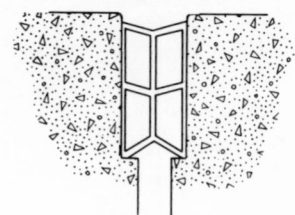

Fig. 19.9 Stepped saw cut with preformed seal.

Large slip-form pavers can pave sections 24 ft wide, and even wider if necessary. Consequently, a longitudinal joint is sawed in the pavement to separate the lanes. This longitudinal joint is, effectively, another contraction joint. Its depth may be the same as the transverse contraction joint, and dowels are spaced along the joint to transfer load.

The contraction joint is intended to accommodate the movement of one slab spacing. The joint is sealed at the top only. If a liquid sealant is being used, a back-up material is placed in the joint under the sealant to maintain the proper shape factor. (See Fig. 19.6.) If a preformed seal is being used no back-up material is needed, but the saw cut may be stepped to seat the seal properly (Fig. 19.9).

19.4.2   *Expansion Joints*

Expansion joints in pavements function in a slightly different fashion from the isolation joints in buildings. The isolation joint in a building separates the structure into two units. Ties or keyways may connect the parts, but the requirements differ from the pavement slab. The expansion joint in the pavement must permit the movement of the pavement in a longitudinal direction, but it must prevent the relative movement of the slabs in either the lateral or the vertical direction. In addition to the forces caused by expansion and contraction, the pavement joint is subject to the constant pounding of automobiles and heavy trucks. Considering a traffic density of 10,000 vehicles per day, which is not unusual, the joint takes quite a beating.

The expansion joint is, of necessity, a formed joint. It is a full-depth joint through the slab and is wider than the contraction joint. Expansion joint width should depend on expansion joint spacing. Practice varies widely from state to state. Many states use only contraction joints and no expansion joints, except at bridge approaches. Other states use varied spacings for the expansion joints. A relatively common practice is to space expansion joints at approximately 1000 ft. Width of the expansion joint varies from ¾ to 2 in. Since the expansion joint is a full-depth joint,

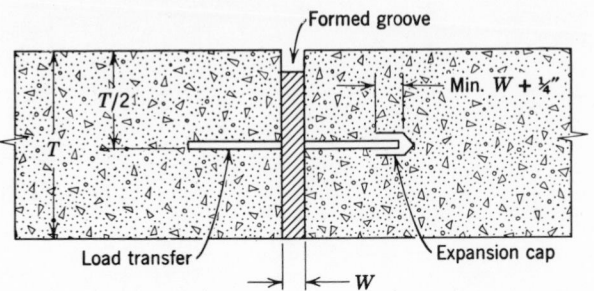

Fig. 19.10 Pavement expansion joint.

a joint filler is usually placed into the joint from the bottom of the slab up to within 1 in. of the top. The expansion joint is then sealed in the same fashion as the contraction joint. The joint fillers used vary widely. Redwood boards, cork boards, and asphalt-impregnated fiberboards are all widely used. The expansion joint, because of its greater width, is generally heavily doweled (Fig. 19.10).

19.5   Joint Forming

The joints in rigid pavements may be made by inserts in the pavement, by sawing, or by hand forming. Hand-formed joints were the first type of joints to be used in pavement construction. However, with the advent of the "paving train" concept of pavement construction, it is not uncommon for paving crews to place more than a mile a day of two-lane pavement. At this rate of construction, hand finishers simply cannot keep up with the rest of the paving operation and, consequently, sawed joints have become the most widely used method of joint construction.

19.5.1   Hand Forming

Before the sawing of joints became practical, hundreds of miles of very good pavement were placed with hand-formed joints. Hand forming is still widely used in smaller paving contracts. In the hand-forming method, the cement finisher simply uses a hand-edging tool to create the weakened plane in the slab. The method does have the inherent disadvantage that the concrete at the joints may be overworked. This overworking results in weak joint faces which contribute to spalling and adhesive sealant failure. However, skilled finishers can do an excellent job of joint forming without overworking the concrete.

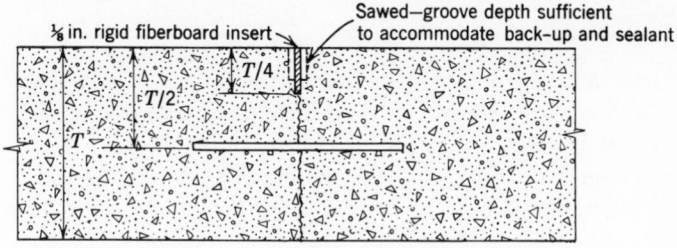

Fig. 19.11 Contraction joint with sawed out insert.

19.5.2 *Inserts*

Inserts have been used to form joints for over 30 years. Dozens of types have been tried, but none have been very widely accepted for any period of time. The two most common types of inserts in current use are the "sawed-out" insert and the "zip-out" insert. The sawed-out insert is a thin sheet of hard fiberboard. It is secured into position before the concrete is placed (Fig. 19.11). The advantage of the sawed insert is that the concrete is allowed to crack below the insert and thus the time of sawing is not so critical. The zip-out insert, generally made of plastic, is fabricated with a tear line. The concrete is allowed to crack and the top portion of the insert is zipped off, leaving the joint ready for sealing (Fig. 19.12).

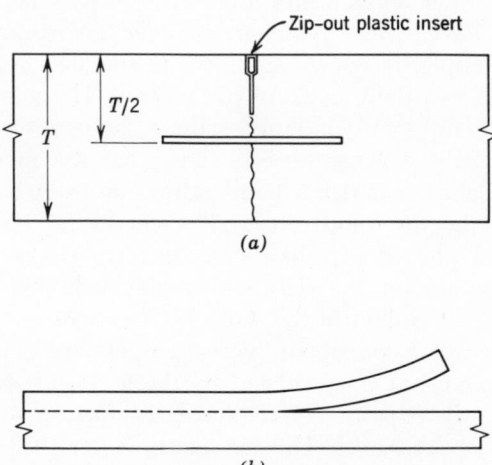

Fig. 19.12 (*a*) Contraction joint with zip-out insert. (*b*) Zip-out plastic insert. Top portion is removed to form groove for sealant; lower portion remains in pavement.

19.5.3   Sawed Joints

The time and manner of pavement joint sawing is extremely important to both joint performance and sealant performance. The time of sawing is related to the curing of the concrete. If sawing is done too soon, the concrete will not have attained sufficient strength. This will result in aggregate pull-out and ragged joints which are difficult to seal. If sawing is delayed too long, the shrinkage cracks will have already formed in the pavement. Another danger with late sawing is that cracks will appear ahead of the saw blade, which results in ragged joints. For normal portland cement concrete, the sawing should be done 48 to 72 hr after placement at 75 F. In hotter weather, the sawing may be done earlier.

Saw cuts in pavements may be formed in several ways. One convenient way to saw contraction joints is to use a single blade in one pass and make the full-depth ($T/6$) cut. A second pass can then be made with a "gang blade," or two blades and a spacer. This second cut saws the joint to its proper width and needs only be deep enough to receive the joint sealant.

19.6   How the Slabs Move

Pavement slabs are generally assumed to expand and contract longitudinally as a function of temperature. Actually, the pavement movement is quite complex. Pavements expand in summer and contract in winter; hence, the joints open in winter and close in summer. Superimposed on this yearly cycle is a daily cycle of movements. The joints open in the cool hours of the night and close under the warm daytime sun. Research studies of the daily pavement cycle show that the movement of the pavement parallels the midslab temperature, and not the slab surface temperature or the air temperature. The curves for temperature and movement, when plotted together, show that the peaks and valleys in the movement curve for the slab are roughly 2 hr behind the corresponding peaks and valleys in the temperature curve.

Pavements do not move smoothly. As temperature goes up, the slab attempts to expand, but is restrained by the friction between slab and subgrade. When the expansion force has built up so that it exceeds the friction, the pavement moves. This movement is fairly rapid, but usually not a sudden lurch of the pavement. Friction in the load transfer device accentuates this "stick and slip" type of movement. If the load transfer system in one joint "freezes" because of misalignment or corrosion, the adjacent joint must accommodate twice the normal amount of movement. This joint freezing will result in sealant failure, regardless of the type of seal being used.

Pavement curl is another factor which complicates the overall movement pattern. Curl is caused by the differential in temperatures between the top and bottom of the slab. Under a hot sun, with the air temperature at 85 to 90 F, the surface temperature of the slab may be as high as 125 to 130 F; whereas the bottom of the slab at the same time may be 75 to 80 F. This temperature differential between the top and bottom of the slab causes the top to expand more than the bottom, so that the slab unit arches or curls upward. In the cool night hours the process is reversed. The top of the slab is cooler than the bottom and thus contracts more, causing the slab to dish upward at the edges. This uneven movement of the slabs resulting from pavement curl may cause the load transfer devices to bind, especially if long slab units are used. The dishing upward of slab edges also causes the slab to have less subgrade support at the joints, which contributes to the cracking of the slab ends and corners under the pounding of heavy truck traffic. Recognition of the problem of pavement curl has caused many states to use shorter joint spacing to minimize the problem.

Moisture absorption also affects the size of concrete slabs. The difference in size between a dry and a wet pavement slab may be as much as the total yearly thermal expansion. However, the pavement is a structure on grade and the moisture content varies from the top to the bottom of the slab. The bottom of the slab is very seldom completely dry. At the present time, there is no accurate method of determining the actual movement due to moisture, and this factor is usually neglected.

Under the action of truck traffic, there is a relative vertical novement of the slab edges. In the warm parts of the day, when the center of the slab unit is arched upward and the slab ends are forced down into the subgrade, the relative movement of slab ends is difficult to measure and may be one or two thousandths of an inch at the most in a relatively new slab. In the very early hours of the morning, however, when the slabs ends are dished upward, the movements are 10 times as great as in the warmer parts of the day. The total movements in a new pavement are still quite small, but there is enough vertical movement under truck traffic to cause fatigue in a sealant. As the slab ages, the vertical working of the slab ends causes the dowels to "pocket," so that the vertical movement of slab ends tends to increase.

The movement of slabs may be calculated by using the temperature range, slab length, and coefficient of expansion for concrete. However, a quick rule of thumb is that highway slabs on grade will move $\frac{1}{16}$ in. for

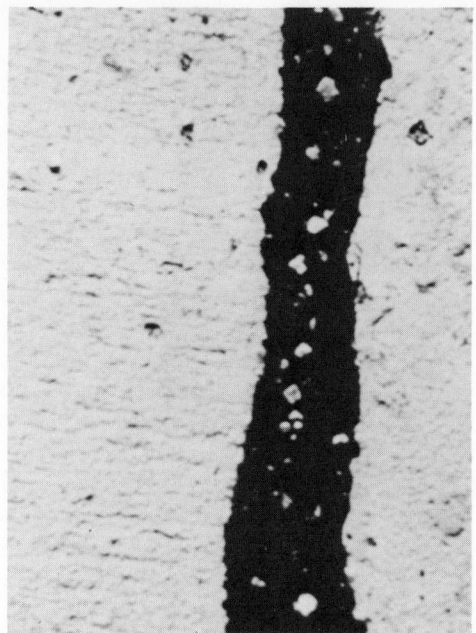

Fig. 19.13 A close-up of a pavement joint, showing stones and dirt embedded into the sealant.

each 10 ft of slab length throughout the temperature range prevailing in most of the continental states.

19.7   What Materials Are Used

Many different materials have been used in the attempt to seal highway pavement joints successfully. The material in the joints must be hard enough to prevent the intrusion of foreign matter into the joint (Fig. 19.13), and should also prevent the passage of water through the joint. In addition, this sealant must have excellent resistance to the deicing salts used on pavements.

The history of paving sealants has been a stormy one. The material which has been used in more joints than any other is the hot poured rubber asphalt sealant. As recently as 1966, 68% of the states still permitted the use of this material. This type of sealant, if well compounded and properly handled in the field, can do a good job of sealing pavement joints.

In the 1950's the polysulfide sealants were introduced to the highway market, to be followed by the urethanes. Unfortunately, the polysulfides were oversold before they were ready for such severe service conditions. The resulting failures caused many state highway engineers to turn a deaf ear to the use of elastomeric sealants for highways. In 1966 only 10% of the state highway departments permitted the use of elastomeric sealants. However, millions of dollars have been spent on research and development of the polysulfides and the urethanes, and some of the newer sealants look very promising.

At the present time, the fastest growing sealants in the highway market are the preformed neoprene gaskets. These compression seals function by exerting a pressure against the faces of the joint. In 1966 only 10% of the state highway departments permitted the use of compression seals. By early 1969 this figure had doubled, and further growth is expected.

The neoprene seals do the best job of keeping incompressible solids out of the pavement joints. They also provide the neatest looking joint of any known seal. The compression seals keep a large percentage of water out of the joints, but are not truly watertight.

The compressions seals, however, have several disadvantages. They cost 5 to 10 times as much as the poured seals. This cost differential may be amortized if the compression seals can function effectively for 15 to 20 years without resealing: only time will tell. The compression seals also require straight, firm joint walls in order to function properly. Furthermore, spalls are difficult to repair. The spalled pavement edge must be rebuilt to its former line and grade, and a new compression seal must be installed.

The presealed joint, consisting of two stainless steel plates and a preformed compression seal, is still in the developmental stage and has not been used to any significant degree. Figure 19.14 shows a test section being installed.

The cold poured bituminous sealants are rapidly disappearing from the highway scene. These materials cure or set by solvent release and have only fair sealing properties. Very few highway officials express any interest in continued use of this material.

19.7.1   Hot Poured Rubber Asphalts

The hot poured asphaltic materials have undoubtedly been the most used and the most misused of any highway joint sealant. The original materials were straight asphalt. In an effort to upgrade the material, finely ground rubber was added. The rubber was obtained from cleaned, ground-up devulcanized used tires. The early rubber asphalt sealants

Fig. 19.14 Installation of a presealed joint.

contained approximately 25% ground rubber and were good quality sealants. As competition increased, the percentage of rubber began to drop, so that by 1967 few rubber asphalt sealants contained as much as 5% rubber. A new and somewhat tighter specification has forced the upgrading of these sealants since 1967.

The hot poured rubber asphalt sealant should be made from good quality asphalt and should contain at least 25% ground rubber. The rubber should be SBR and should be uniformly ground to a 30 mesh size. Hot poured rubber asphalts, using neoprene and other synthetic rubbers, are currently being manufactured. A great deal of both laboratory and field testing remains to be done, but the neoprene asphalts at this time look very promising.

The hot poured sealant must be carefully handled at the job site. The sealant is furnished to the job site in solid form in large drums. The material must be heated uniformly to a melt, so that it can be extruded into the joint. The heating should be done in a jacketed, oil-bath heating kettle. Temperature must be carefully maintained. A temperature of 400 F can safely be tolerated, whereas temperatures over 450 F will "burn out" the material so that it will not function properly. Also, these materials should not be reheated from one day's work to the next.

The hot poured sealants, when properly placed, can effectively seal joints for a period of 5 years. These sealants are rubbery materials and should be placed into clean, sound joints with a reasonable shape factor. The optimum 1:1 width to depth ratio in the joint should be used whenever possible. However, these are low recovery sealants and the depth of the joint may be greater than the width, if necessary, in order to obtain sufficient adhesive area. In the normal ⅜-in.-wide contraction joint, the depth of seal should be ½ in. This shape factor can be obtained by the use of butyl rod stock and a bond breaker placed into the joint before installation of the sealant.

The hot pours have good adhesion to concrete without the use of primers. These sealants have only fair recovery properties, but good resistance to deicing chemicals. They are especially useful for resealing work, and for crack sealing in both rigid and flexible pavements. The hot pours are also useful for sealing badly spalled joints in older pavements. In this application both the preformed seals and the elastomers are useless.

A hot poured sealant can be installed in a pavement joint for approximately 20 to 30 cents per foot of joint. At this price, even if resealing is required every five years, the sealant is still a good buy. This material also has the advantage of familiarity. Construction and maintenance crews know the material and the equipment necessary to install it. Many state highway crews and contractors now have modern truck- or trailer-mounted equipment capable of sealing 20 ft of contraction joint per minute.

19.7.2 Cold Poured Elastomers

The only elastomeric sealants used in highway joints are the polysulfides, urethanes, and polymercaptans. The polysulfides were the first to be used and are still the largest sellers among the elastomers. Within the past five years the urethane sealants have grown rapidly, and are now almost as widely used as the polysulfides. The polymercaptans have been formulated for highway work, but have not yet been used to any significant degree.

The elastomers have not been notably successful in sealing highway joints. Many states have tried test installations of these polymeric sealants, but only about 10% of the states have accepted the elastomers as a specification item. The elastomeric sealants got off to an unfortunate start. The materials looked good in the laboratory, but usually failed in the field. Many of the early installations were placed in poorly prepared joints, and in some instances the two component materials were not properly mixed. Even under the best of conditions failures were frequent.

The polymeric sealants which first appeared on the market were not properly formulated as paving sealants. The elastomeric sealants cost approximately 40 to 60 cents per foot of joint, which is roughly twice the cost of the hot poured rubber asphalts. Consequently, many of the early sealants were compounded with one eye on the price, and performance suffered badly.

The price per performance picture is beginning to change. The state highway officials are thinking more in terms of cost per year over an extended time period. They are willing to pay a higher initial cost if higher performance accompanies it. Consequently, since 1966 some manufacturers have produced and marketed high quality polysulfides and urethanes which look very promising. However, failures are long remembered, while the successful installations often go unnoticed. Because of the early failures, the elastomeric sealants lost a major segment of the market to the preformed compression seals. This loss may never be regained.

The elastomeric sealant for use in highway joints should be a two-component material. The one-component materials cure too slowly to be practical. The sealant should have a Shore hardness of 15 to 25 (at 70 F) in order to resist penetration by stones and road dirt. The sealant should have tensile adhesion to concrete of at least 100 psi. The sealant must also have an ultimate elongation of at least 150% (in the $\frac{1}{2} \times \frac{1}{2}$-in. specimen shape). The normal $\frac{3}{8}$-in.-wide contraction joint in a 60 ft slab may open more than 100% from summer to winter; hence this elongation requirement cannot be compromised. The sealant should also be of the pourable or "self-leveling" type in order to fill all voids in the substrate without being tooled after installation.

The sealants in pavement joints operate under extremely severe service conditions. Consequently, these materials must be placed with great attention to detail. The joint opening must be properly cleaned and primed, and a proper shape factor must be maintained. The depth of the sealant material should never exceed its width. In the normal $\frac{3}{8}$-in. contraction joint, the depth of the sealant should be $\frac{3}{8}$ in. In expansion joints, which accommodate larger movements, the depth of sealant should be $\frac{1}{2}$ the joint width, if possible.

The two-component elastomers for highway use may be either hand-mix or machine-mix materials. The hand-mix materials are furnished in

Fig. 19.15 Trailer-mounted sealing machine for two-component elastomers. (Courtesy of Allied Materials Corporation)

the usual ratio of 15 parts curing agent to 100 parts polymer. These materials are mixed by an agitator blade on a slow speed electric drill and may be either poured or gun-extruded into the joint. The machine-mix materials are furnished in a 1:1 volume ratio. The two components are placed in separate chambers on the mixing machine, pumped through separate hoses, and intimately mixed at the nozzle. Then they are extruded under pressure into the joint. The machine is usually truck or trailer mounted and can mix and install 20 to 25 ft of contraction joint per minute. Figure 19.15 shows a typical trailer-mounted machine mixer.

19.7.3 Preformed Compression Seals

The preformed compression seals are the fastest growing type of highway paving seal. Although some experimental sections of extruded silicone and EPT have been used, virtually all the compression seals in current use are made of neoprene.

There are hundreds of different shapes of seals currently available, but the most frequently used sections are the chevron and the rectangular shape (Fig. 19.16). Although no comparative stress analysis has been

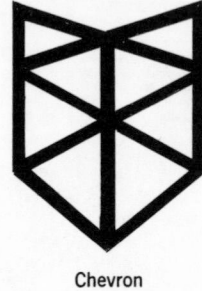 

Chevron   Square

Fig. 19.16 Preformed seal cross sections.

published, the rectangular shape is believed to have the better properties.

The compression seals must be carefully selected and installed, in order to function properly. A compression seal should be compressed to approximately ½ of its initial width for installation, so that with the joint open to its greatest width the seal will continue to exert pressure against the joint wall. The following example demonstrates the selection of such a seal based on certain installation qualifications.

> Joint spacing 50 ft
> Width of contraction joints ⅜ in.
> Temperature range −20 F to +120 F
> Seal installed at 70 F

$$\text{Joint movement} = 50 \text{ ft} \times 12 \text{ in./ft} \times 0.0000065 \, \frac{\text{in./in.}}{\text{deg F}} \times 90 \text{ F}$$

$$= 0.35 \text{ in.} \simeq ⅜ \text{ in.}$$

Maximum joint width = ⅜ + ⅜ = ¾ in.

Select a seal ¹³⁄₁₆ in. wide.

This seal can be compressed to slightly less than ⅜ in. for installation and will continue to exert a sealing pressure when the joint is at its maximum width. Figure 19.17 shows the seal in the three stages at which it must function.

The compression seals must be properly installed in order to function properly. The seal must not be stretched during installation. A seal which is stretched longitudinally during installation will neck down and make installation easier, but will not perform properly after it is in the joint. For example, a ⅞-in.-wide seal, when stretched, may neck down to ¾ in. wide. In effect, then, a ¾-in.-wide seal is being placed into the

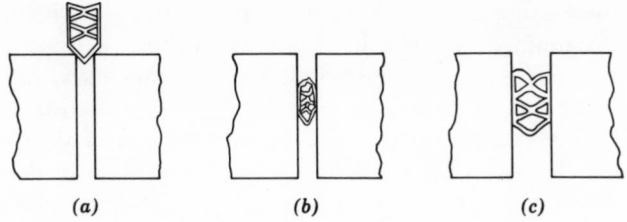

(a)   (b)   (c)

Fig. 19.17 Preformed neoprene (*a*) before installation, (*b*) installed in ⅜-in. joint, and (*c*) after joint opens to ¾ in. in cold weather.

joint. In the sample problem just discussed, a ¾-in. seal would obviously not be effective when the joint is at its maximum width.

In the early days of compression sealing, the seals were placed with a flanged hand roller. However, the roller was slow and produced too much stretch in the seal. Currently, there are several power operated machines available which precompress the seal, place the lubricant-ad-

Fig. 19.18 Installation of a preformed compression seal. (Courtesy of D. S. Brown Corporation)

hesive on the joint wall, and insert the seal into the joint at the proper depth with a minimum of stretch. Figure 19.18 shows a typical seal installation machine, which can install 20 ft of preformed seal per minute.

The preformed seals do have a cost disadvantage. It costs $1-$3 per foot to seal a pavement joint in new work, and the cost may go to $4 per foot or higher for resealing work. However, if the preformed seals can function effectively over a 20-year period, the additional cost may be justified. Consequently, *if* the compression seals can do the best job of protecting the pavement from distress over an extended time span, they will be the highway engineer's best sealant investment.

The compression seals also have other disadvantages. In order to function properly, the preformed compression seal must be installed in joints with straight, firm joint walls. Also, although the compression seals do the best job of keeping incompressible solids out of the joint, they are not watertight. Consequently, water seeping past the compression seal can corrode the load transfer device and thus inhibit normal pavement movement. As with any other type of seal, the preformed seal depends on each slab unit to accommodate its share of the movement. Another difficulty with preformed seals is related to the forming of the joints. The seals for any given project are designed for the joint size. However, after the joints are sawed, it is often found that some of the joints have cracked through, whereas others have not. Consequently, the joints may not be uniform in size at the time of installation.

19.8 Specifications

Joint sealants for highway construction are specified by the highway engineers. There are approximately 75 state highway departments or turnpike authorities and over 600 city and county authorities which have some jurisdiction over pavement construction. Every state highway department has its own set of contruction specifications and many of the city and county authorities have similar codes.

The sealant specifications included in most of these construction codes are modeled after applicable ASTM or Federal specifications. The following specifications have been used, in whole or in part, by highway engineers in drafting the state or county codes.

Designation	Description
ASTM D1850–61T	One- or two-component (no composition restriction)
SS–S–158a	Asphaltic; solvent type
SS–S–156	Emulsion type (no composition restriction)

SS–S–170	Two-component; synthetic rubber; jet fuel resistant
SS–S–00195	Two-component; polymer type; machine mixed
SS–S–159b	Multiple component; mastic type
SS–S–00200c	Two-component; polymer type; jet fuel resistant; machine mixed
ASTM D1852–61T	Two-component; polymer type; jet fuel resistant
SS–S–00164	Asphalt, with or without rubber
SS–S–001401	Asphalt, with or without rubber
ASTM D1190–64	Elastic type (no material restriction)
ASTM D1864–61T	Elastic type; jet fuel resistant
SS–S–171	Asphalt, mineral filled
SS–S–00167b	Jet fuel resistant (no material restriction)
ASTM C509	Cellular elastomeric preformed gasket
AASHO–ARBA *	Preformed elastomeric compression seal
ASTM D1752–60T	Sponge rubber, cork, and self-expanding cork
ASTM D1751–65	Cork or cane fiber, with felt and bituminous binder
ASTM D545	Closed-cell organic foam

Specification TT-S-00227 for two-component building sealants is also used as a guide specification. From this maze of specifications, a few facts stand out. The acceptance procedure for Federal specifications must be kept in mind. Any manufacturer may test his own material according to the specification requirements and may state in his advertising that the material meets or exceeds a given Federal specification. Fortunately, most sealant formulators are in business to stay, and value their reputations. However, it should be apparent that it would be quite simple for a fly-by-night compounder to select the recovery requirement from one specification, the bond test requirement from a second specification, and the elongation requirement from yet a third specification. This compounder could advertise a pretty convincing case for a sealant that was actually useless.

Many state highway departments now have very well equipped laboratories for acceptance testing, and most of the states will insist on laboratory verification or actual field tests before accepting a new sealant material. Most of the state highway departments are quick to test new materials, but very conservative about actual acceptance. The city and

* American Road Builders Association.

county jurisdictions usually have competent personnel, but are not equipped with testing laboratories. Consequently, they must rely on the guide specifications or leadership from the state highway departments.

From the long list of specifications given above, three or four stand out as being of any significance. For the hot poured asphaltic materials, SS-S-00164 and SS-S-001401 are applicable. The 164 specification had, for years, been the standard specification. The 1401 specification is a recent amendment. The addition of a resilience requirement forces the manufacturer to upgrade the rubber content of the material. The materials which meet this newer specification look good, but require long term field verification.

In the field of elastomeric sealants, a good guide specification simply does not exist. The applicable specifications are SS-S-00195 and SS-S-00200. These two are known throughout the trade as "coal tar specs," although the only material restriction in the specification is that the sealant be formulated from an elastomer. The original versions of these specifications were written around 1959. They incorporated some of the basic ideas from the hot poured sealant specifications, as well as the best technology of elastomers that was available at that time. Both of these specifications are now in need of a complete revision. Sealants meeting the laboratory requirements of these specifications have had moderate success in airfield pavements, but have had a poor performance record in highway pavement joints. Some states have tried writing their own specifications using the building sealant specification as a guide, and have fared slightly better. Fortunately, the manufacturers are aware of the size of the market, and are aware, also, that they must convince the materials engineer of the state highway department if they wish to sell. Consequently, the manufacturers, rather than the specification groups, are pushing the research and development of better elastomers.

The specifications for the preformed compression seals were, for many years, written around a tentative specification published by E.I. DuPont de Nemours, Inc. The DuPont specification in general was very well written and, with slight amendment, has been adopted by many states. The AASHO–ARBA specification consists of the basic DuPont specification with a few modifications. NCHRP has funded a very large contract on compression seals, and a definitive specification for the preformed seals should be available for adoption by 1971.

In the listing of specifications given above, a number of specifications list resistance to jet fuel as a requirement. These specifications were obviously written for airport pavement work and have been adopted into the highway field.

19.9   Crack Sealing

Cracks occur in both asphaltic concrete and portland cement concrete. The sealing of the cracks may vary depending on the type of pavement and the size of the cracks.

Cracks in bituminous concrete may be either reflection cracks or random cracks. Reflection cracks are cracks in bituminous overlays placed on portland cement pavements. These cracks are fairly straight and correspond to the joint spacing in the pavement below. Random cracks may be caused by thermal change, frost heave, subgrade settlement, or any combination of these factors.

Properly sealing the cracks in bituminous pavements is a more difficult job than sealing the joints in rigid pavements. Because of the irregularity of the cracks, it is impossible to use a preformed compression seal. Elastomeric sealants are useless because there is no sound joint interface to which the sealant can adhere; also, most of the elastomers are incompatible with asphalt and, consequently, will not bond. Tar and asphalt have thus been the two materials used to seal these cracks. The tar is applied hot. The asphalts may be either a hot poured or a cold poured solvent release type. The hot poured asphaltic materials are the best choice. Cracks have sometimes been widened with a small routing tool to create a better bond interface, but the routing has not made any notable change in the sealant performance.

Sealing of large areas of random cracks (map cracking) has been accomplished by various types of seal coatings. The seal coatings, if properly applied to a clean pavement, may be effective for as long as five years.

The cracking of rigid pavements may take several forms, but the most prevalent is the transverse crack which is caused when the normal contraction joint is immobilized. Cracks in rigid pavements may be repaired structurally, or they may be simply sealed. The structural repair is accomplished with an epoxy adhesive which may be either poured or pumped into the crack. Structural repairs look good in theory, but have not been in use long enough to judge their effectiveness. Crack sealing in rigid pavements is subject to many of the same limitations that apply in bituminous pavements. The cracks are too irregular for compression seals. Also, the cracks are characterized by "aggregate pull-out" and weak interfaces, and thus elastomers cannot be used. Consequently, tars and asphalts are used, as with the bituminous pavements.

(a)

Fig. 19.19 Complete sequence of resealing an airport pavement. (*a*) Routing the old sealant from the joint. (*b*) Cleaning the joint wall by sandblasting. (*c*) Vacuum-cleaning the joints. (*d*) Priming the joints. (*e*) Installing the back-up material. (*f*) Installing the sealant. (Courtesy of Products Research and Chemical Company)

19.10   Sealants in Airport Paving

The problem of sealing airport pavement joints is somewhat different from that of sealing highway pavement. This difference in sealing practice is mainly due to the difference in slab depth and the difference in panel size. In highway construction, the average slab thickness is approximately 9 in. In airport construction, the slab thickness may be twice this amount, or even more, because of the heavier wheel loads which the pavement has to support. Consequently, the airport pavement is slower to respond to temperature changes than the highway slab. This difference in slab thickness also demands a different type of joint con-

(b)

(c)

193

(d)

struction. The ordinary saw cut contraction joint is usually not applicable to airport paving.

In highway construction using a contraction joint spacing of 50 ft, the highway is generally paved in a two-lane width and then the joints are sawed. Contraction joints are placed every 50 ft, and a longitudinal joint is sawed which divides the pavement into two 12-ft lanes. The pavement slab is thus subdivided into panels 50 ft long by 12 ft wide. In airport construction, especially in parking and ramp areas, the pavement is usually placed in square panels which are approximately 25 ft on each side. There is no real differentiation between longitudinal and transverse joints. Airport runways, of course, are several thousand feet long and about 100 ft wide, so that transverse and longitudinal joints can be identified; however, the construction is still usually in the form of the square panels.

Because of the relatively short joint spacing and greater slab thickness, there is little or no curl in airport pavements. Almost all of the movement which occurs in airport pavements can be attributed to temperature and moisture changes.

Joints in airport pavements are generally formed with some form of insert. The insert may be a plate which is removed as the finisher forms the joint by hand, or it may be a sawed-out fiberboard insert. The sawed-out insert has become the more popular method of joint forming.

The materials used in airport pavements are much the same as those used in highway construction, except for the addition of a requirement for jet fuel resistance. The sealant and its adhesive bond to the concrete pavement must not be affected by jet fuel spillage. This is especially critical, of course, in parking and ramp areas where the aircraft are serviced. Specification SS-S-00167, although it contains no material restrictions, is the specification used for the hot poured sealant. Composition of this material differs from the highway sealant since asphalt is not compatible with jet fuel. Specification SS-S-00200 is the usual specification used for elastomeric sealants. For years this has been a "coal tar polysulfide specification," although recently several coal tar urethanes have been submitted to government laboratories for evaluation. Preformed compression seals have been used to a limited extent in airport work,

(e)

(f)

but their growth in this area has been much slower than in highway sealing. Figure 19.19 shows the complete sequence of resealing with a two-component elastomeric sealant.

Actually, the construction of airports can be divided into two major areas: civilian and military construction. In military construction, the Federal specifications apply directly. In civilian airport construction, the work is subject to the jurisdiction of the Federal Aviation Authority. The FAA has its own set of specifications and recommended practices, but the sealant specifications are, for the most part, modeled after the Federal specifications.

One additional sealant requirement is brought about by the difference between civilian and military aircraft. On almost all civilian aircraft, such as the 707, the 727, and the DC–8, the jet engines are horizontal, so that there is usually no direct blast of jet exhaust down against the pavement. Many military aircraft, on the other hand, have the jet engines mounted at a slight angle of incidence, so that there is a cone of jet blast directed toward the pavement behind the aircraft. Consequently, the thermoplastic, or hot poured, materials are not the best choice for military construction.

20

Sealants in Bridge Expansion Joints

20.1 Introduction

Bridges are harder to seal properly than pavement joints. Bridges are subject to all the forces that move pavements, and a few more besides. In addition to being harder to seal, bridge joints are more critical than pavement joints. Poorly sealed bridge joints, as well as being very expensive to repair, can result in conditions which are very hazardous to the motorist.

For drainage purposes, the bridge is usually on at least a minimum grade and may also be crowned. Consequently, the unsealed or poorly sealed expansion joint simply serves as a funnel to direct the flow of water and de-icer solutions on to bridge seats and pier caps. In ordinary pavement joints, incompressibles in the joint are the major problem; however, in bridge joints the most serious problem is water leakage.

20.2 History

In the past, bridge joints have been handled by a wide variety of methods. Sliding plates and finger joints were in use 30 years ago in order to bridge the open joint. Many states used gutters and downspouts to carry off drainage water. The first sealing materials to be used to close the expansion joints were the hot poured asphalts. In the 1950's the polysulfides were first used as bridge joint sealants. In the early 1960's the preformed neoprene seals were first introduced into bridge sealing. Bridge joints are usually much wider than pavement joints, and larger preformed seals were therefore developed for this application. In recent years, the trend toward longer spans and continuous construction has put greater demands on the expansion joint sealant. Thus the single pre-formed compression seal evolved into multiple units. Various patented systems, such as the Transflex * bridge seal, have been developed in this country. Likewise, engineers in some of the European countries have developed very sophisticated joint systems which consist of pre-formed seal units, prestressed for installation at a particular tempera-ture.

20.3   Why Seal the Joints?

Expansion joints in bridges are sealed primarily to keep out water. The deicing salts used on the bridges and pavements in winter are excellent for their purpose, but are destructive to concrete. Water and de-icer solu-tions passing through bridge joints corrode both the ends of steel beams and the roller bearings on which the bridge rests. This water also carries with it dirt and debris which, in addition to staining pier caps and col-umns, may inhibit the normal action of roller or sliding bearings. Water washing down through the joint and over the bridge seat also contrib-utes to the migration of back slopes. In colder climates, water and ici-cles may drop onto the roadway or traffic below. The most severe dan-ger, however, is the heavy flow and concentration of de-icer solutions on to bridge seats and pier caps. These are shaded areas and the de-icers have ample time to saturate the concrete for maximum destructive effect. All in all, many state bridge engineers are in agreement that it is easier to try to seal the joint than to maintain an open joint.

Some of the results of poor joint sealing are shown in Figs. 20.1 and 20.2. Structural members which are subject to attack have deteriorated in a remarkably short time. Figure 20.1 shows a pier cap which has begun to deteriorate. Figure 20.2 shows a bridge column which is badly disintegrated. This structure was approximately 10 years old when the photograph was taken.

20.4   Structure of the Joint

Bridge expansion joints more nearly resemble the isolation joints in buildings than the expansion joints in pavements. They are "through-

* Trademark of General Tire and Rubber Company.

Fig. 20.1 Pier cap which has begun to deteriorate due to deicing chemicals.

the-structure" joints that completely separate the bridge from the approach pavement. The structure of the joint will vary depending on the span length and type of construction. Figure 20.3 shows a typical expansion joint in a steel beam bridge. This type of joint is used on relatively short spans and places the joint seal into a butt joint, so that the seal is subjected to a tension-compression movement cycle. Figure 20.4 shows a sliding plate joint which is also applicable to short spans. The sliding plate joint is usually not sealed and, consequently, is not watertight. The construction of a finger plate joint is shown in Fig. 20.5. For bridges of longer spans, such as cantilever and suspension bridges which exhibit very large movements, individual bearings systems and sealing systems are usually designed. Figure 20.6 shows a steel reinforced, preformed elastomeric seal.

20.5   How the Bridges Move

As with most other structures, the movement of bridges is assumed to be due to thermal expansion and contraction. However, the deflection of

Fig. 20.2 Deteriorated bridge column.

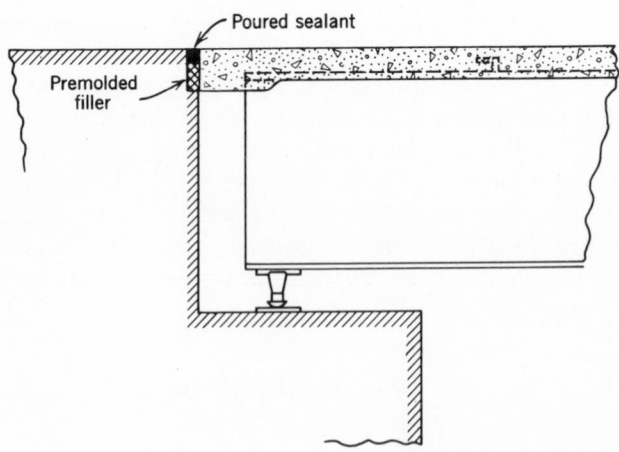

Fig. 20.3 Expansion joint in a short-span steel beam bridge.

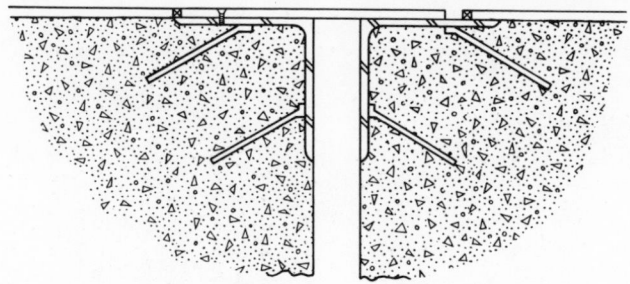

Fig. 20.4 Sliding plate joint.

the bridge under live load is also responsible for joint movement. The midspan deflection of a bridge under traffic causes an end rotation of the structure which opens the expansion joint. Simple computations show that in a 60-ft steel beam bridge with a concrete deck, the expansion joint opens by an amount equal to ½ the yearly thermal movement of the bridge every time a 20-ton truck passes over the structure. These computations have been checked by the electronic instrumentation of many structures. In winter months, when the bridge contracts, this deflection movement is added to the movement caused by contraction of the structure. The trend in bridge design is toward longer, more flexible structures and thus the deflection movements should not be ignored.

The braking action of traffic on bridges is also responsible for bridge movement on short-span structures. In a new bridge, this longitudinal thrust of the bridge puts a dynamic load of tension on the seal at one

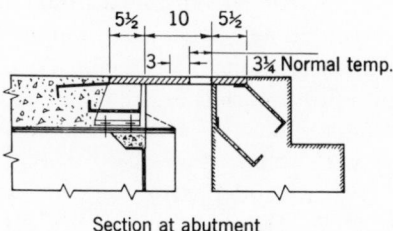

Section at abutment

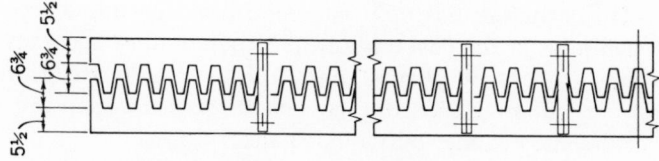

Fig. 20.5 Finger plate expansion joint.

Fig. 20.6 Steel reinforced, preformed elastomeric seal. (Courtesy of General Tire and Rubber Company)

end, and compression at the other end. The actual amount of this movement is unknown because it is so difficult to separate it from other bridge movements, even with electronic instrumentation. It is known, however, that as the bridge ages, there is a longitudinal shoving of the structure in the direction of traffic. This factor alone could cause the resealing of a structure after a period of a few years.

Continuous construction has also served to increase joint movements. A three-span bridge with spans of 50-80-50 ft might be constructed as three simple spans with a fixed bearing and an expansion bearing on each pier cap (Fig. 20.7). However, many bridges in this same span range are now built as three-span continuous structures. With this type of construction, the support bearings might be arranged as shown in Fig. 20.8. This structure has only one fixed bearing and, consequently, the expansion joint at the east end of the structure must accommodate a total movement due to 80 plus 50 (or 130) ft of span length. In the simple beam bridge of Fig. 20.7, the movements for each span are accommodated at the ends of the individual spans.

Bridges are also known to be more responsive to thermal changes

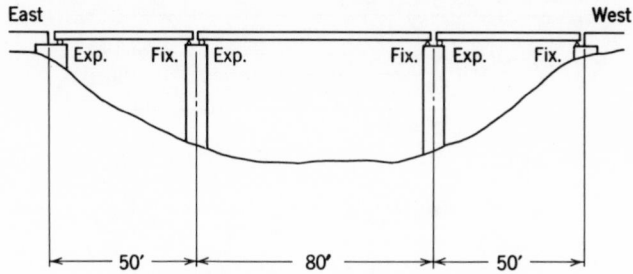

Fig. 20.7 Three-span bridge built as simple spans.

than are pavements on grade. The highway pavement surface is subject to great thermal changes, but the bottom of the pavement slab is fairly well insulated from these thermal fluctuations. The bridge structure, on the other hand, is exposed to temperature changes under the bridge as well as at the top surface. There is some differential due to the shading of the underside, but the overall effect is that bridges respond to temperature changes much faster than do pavements on grade.

Another factor which must be considered in the design of bridge joints is the skew of the bridge structure. By far, the larger number of bridges are installed at some angle of skew. Figure 20.9 shows a plan view of a typical skew bridge. In the skew bridge, the main structural members are aligned with the direction of traffic, so that the thermal expansion and contraction is in this direction. The effect of the skew is to place the joint seal in longitudinal shear as well as tension and compression. This combination stress makes the joint very difficult to seal. Preformed seals tend to bunch up and "walk" out of the joint. Poured sealants are subject to a peeling action which tends to break down the adhesive bond of the sealants to the concrete.

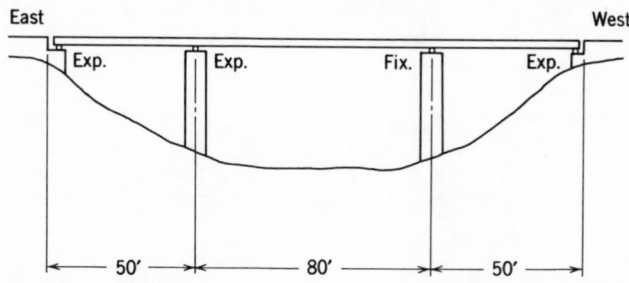

Fig. 20.8 Three-span continuous bridge.

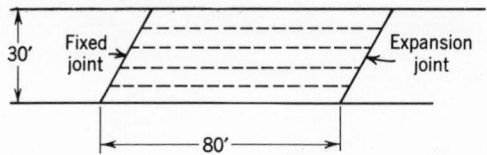

Fig. 20.9 Plan view of a single span skew bridge.

20.6   *What Materials Are Used*

The hot poured asphalt sealants are not well suited to the dynamic movement of bridge joints, but are still used to some extent in short-span bridges. These materials do not have enough resiliency and elongation capacity to cope with the movement of bridge structures.

The elastomeric sealants, when first introduced, were used in a great many bridge joints, without much success. Since further research has shown the extent of bridge movements, the manufacturers of elastomeric sealants have confined their efforts to the shorter span structures and have fared much better.

The elastomeric sealant for use in bridge joints should have approximately the same properties as the pavement sealants. Although incompressibles are not the most serious problem in bridge joints, the sealant should have a hardness of 15 to 25 at 70 F to prevent puncture by sharp stones and grit. The elongation requirement of 150% is especially critical in bridge joints. The bridge sealant should also have a good recovery of at least 75% after being blocked into a strained position for 24 hr. Most of the urethanes and some of the higher recovery polysulfides will meet this requirement. Above all, the sealant must have excellent adhesion to both concrete and steel. In general, if the calculated movement of the bridge joint exceeds ½ in., *do not* use an elastomeric sealant.

When an elastomer is to be used for sealing bridge joints, the bridge designer has the opportunity to use the 2:1 width to depth ratio in the

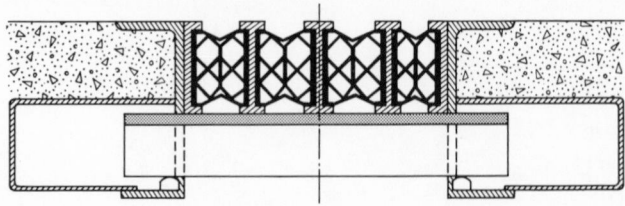

Fig. 20.10 Modular preformed sealing unit. (Courtesy of Acme Highway Products Company)

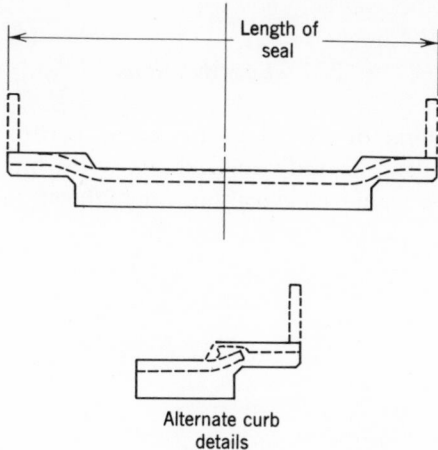

Fig. 20.11 Curb details for preformed bridge seals.

sealant material. Installation procedures outlined in Chapter 6, concerning back-up materials, bond breakers, and joint cleaning, are especially critical in bridge joints. The elastomeric sealants are improving and are now being accepted by some states for use in bridge joints. For example, California has recently approved the use of a urethane sealant for use in short-span bridges with small joint movements. The elastomers will probably take over the market for bridge sealing in short-span bridges, because the preformed seals are expensive and are not watertight. However, in order to accomplish this take-over, better elastomers and better installation practices will have to be developed.

The preformed compression seals in single and multiple units have taken over virtually the entire sealing market of medium- and long-span structures. Figure 20.10 shows a multiple, or modular, sealing unit to accommodate 12 in. of movement. As of this time, most of the seals are made of neoprene, although EPT and other rubbers are being developed. The preformed seals do keep a large percentage of the water out of the joint, but they are not watertight. Installation details are particularly important if the joint is to be as dry as possible. The compression seals cannot be successfully bent to small radii. This creates special problems at the curb line because the seal tends to "bunch up" if it is bent from the bridge deck up the face of the curb. The problem is compounded because the bridge deck may be crowned, thus causing the greatest flow of water at the curb line, where the seal is least effective. Figure 20.11 shows a detail which has been reasonably successful in sealing the joint at the curb line.

20.7 Specifications

There are no Federal or ASTM specifications written specifically for bridge joint sealants. The individual states write their own specifications, using the applicable specifications for highway joints as a guide.

21

Adhesives

21.1 Introduction

Adhesives are here to stay. History shows us that adhesives have been used since the time of the early Egyptians and Assyrians. The Egyptians used adhesives for the bonding of papyrus, and pieces of veneered furniture have been found in the tombs of some of the pharaohs. These adhesives were undoubtedly of animal origin. The Assyrians not only contributed mud brick, our earliest known molded building material, but in many cases bonded the brick together with a cementitious mortar based on clay. These same Assyrians discovered bitumen springs near the Euphrates River and used this pitch as a cementing material. The Greeks were the pioneers of faced and veneered construction. They often built structures of limestone which were veneered with a fine grained marble. The marble was sometimes attached with a mortar adhesive. The early Romans also used a faced concrete type of construction. The Chinese used mortars in their structural work and animal glues in their decorative building work.

In the Western world, mortars, bitumens, and starches have been used freely, but the growth in the use of adhesives, up until 30 years ago, has been pretty much the growth in the use of animal glues. Within the last three decades an explosion has taken place in the number and types of materials available for adhesive bonding: epoxies, polyesters, polyvinyl acetates, phenolics—the list seems almost endless and is still growing.

It looks almost impossible to discuss in detail all the adhesives in use in the construction industry today. However, this is not really necessary. On a value basis, approximately 60% of the adhesives manufactured are for captive consumption; for example, the company that manufactures both the adhesive and the finished fiberboard, or the company that manufactures both the adhesive and the finished sandpaper. In this book the captive adhesives will be touched on, but the emphasis will be placed on those materials which the architect or engineer can specify for use in construction.

21.2 The Market

Accurate estimates of adhesive consumption by the construction industry are hard to obtain because these materials are sold under such names as glues, mortars, adhesives, mucilages, and resins. It has been estimated that over 300 companies manufacture and market adhesives under brand names. A similar number of companies manufacture adhesives solely for captive use. Some of the larger companies channel their output to both captive and outside use, but these are relatively few in number. The picture that emerges is of an industry dominated not by a few large producers, but rather by a cluster of smaller companies all competing for a share of the market. Plants tend to be small and widely scattered geographically to serve specific outlets. Even the larger companies tend to have smaller plants, scattered to give good geographical coverage. The effect of this industry distribution on the construction business is that quality control on an industrywide basis is hard to achieve. Also, specifications are harder to write and enforce than in other areas, such as steel construction.

Table 21.1 is an estimate of the major uses of adhesives in the construction industry in 1966.

This table does not attempt to list all the adhesives used for the various applications. The compositions shown are simply representative of a given group and may not be the best adhesives available. Subsequent chapters will list advantages and disadvantages of the various adhesives within each group.

More and more uses are being found every day for adhesives in construction. Adhesives for concrete are growing rapidly, and so are the pressure sensitive adhesive tapes. Metal-to-metal bonding, which not too long ago was considered impossible, is now being done with nonstructural components and is being investigated for structural joints. This widening of the field of application of adhesive bonding has been due mainly to the growth of the thermosetting resins, such as the epoxies and polyesters. These synthetic resin adhesives are often desirable because they behave better in moist environments than do animal or vege-

Table 21.1

Type	Composition	Volume	Usage
Roofing cements	Asphalt and coal tar pitch	370 million lb	Specified by architect
Floor and wall tile adhesives	Asphalt, rubber, polyvinyl acetate	120 million lb	Specified by architect
Thermal insulation binders	Phenolic resins	115 million lb	Captive
Concrete adhesives	Epoxy resins	2 million lb	Specified by engineer
Plywood particle board glued laminated members	Urea resin, casein resorcinol	350 million lb	Captive
Wallpaper pastes	Starch	10 million lb	Specified by architect
Pressure sensitive adhesive tapes	Variable	Unknown	Specified

table glues. Over the past decade these synthetic resins have grown at a rate of approximately 10% per year, whereas the overall growth rate for adhesives has been about 4% per year.

21.3 Why Adhesives Are Used

Adhesive bonding has many advantages to offer the construction industry. No other method of attachment is satisfactory for many applications. It would be absurd to consider nailing a ceramic wall tile into position or using a plywood paneling which had the wood plies stapled together. Even the piece of sandpaper in the carpenter's tool box depends on an adhesive to hold the grit to its paper backing. When all the applications of adhesives are taken into account, adhesive bonding must be considered as the most widely used method of holding materials together.

Mechanical fasteners are, by their nature, discontinuous and thus cause stress concentrations. Even welds, which are considered by many to be the best method of attaching metals, are most often fillet or plug welds and, consequently, furnish only edge attachment. The adhesively bonded joint, on the other hand, furnishes a full film of adhesive over the bonded parts, which results in a more uniform stress distribution.

Adhesive bonding in many cases offers the contractor easier installation, together with the resulting savings in construction dollars. Adhesively bonded wallboard requires less plastering and sanding. Acoustical tile ceilings can often be installed faster by bonding than by stapling. Insulation batts attached to wall studs can be installed faster and with a better seal by using adhesives.

21.4 Disadvantages of Adhesive Bonding

The greatest disadvantage of adhesive bonding is uncertainty. No adequate nondestructive means of testing an adhesive bond is presently available. In fully structural applications this may well mean that the particular component is never really tested under full service load until it is placed in the structure. It may then be too late.

Properties of the actual glue line are somewhat difficult to determine. The modulus of elasticity of the bonding layer should ideally approach the modulus value of the substrate. This matching becomes extremely difficult, especially in metal bonding. In the bonding of dissimilar parts, such as wood to metal, different coefficients of expansion of the substrates may be a problem. The coefficient of expansion of the bonding layer itself may be a problem.

21.5 Where Adhesives Are Used

Adhesives are used in almost every phase of the construction industry. The use of adhesives in construction began with the finishing trades. Flooring materials, wallpaper, and roofing cements were the first volume applications. Other wall coverings, such as tile and paneling and then ceiling panels, soon followed. Adhesives have spread into both semi-structural applications, such as the placing of gypsum wallboard, and purely structural uses, such as the glued–nailed joints in wood trusses and in the lamination of large timber girders and frames.

21.6 The Nature of Adhesion

The total adhesive force holding two materials together is the sum of two factors, namely, specific adhesion and mechanical adhesion. Specific adhesion is chemical. It is the molecular attraction between two materials. The actual bond may be a chemical union, such as the sulfur linkage in rubber-to-metal bonding, or it may be simply an electrical attraction between electrons of the two substances. Mechanical adhesion is the bonding force provided by interlocking action. Specific adhesion can be considered as an active force holding the materials together. It is effective under tensile-, shear-, and peel-type loadings; whereas mechanical adhesion is passive and not very effective until acted upon by an outside force. Mechanical adhesion is most effective under shear-type loading and contributes little to the tensile strength of a joint.

The total force holding two materials together is proportional to the bond area. It is a common misconception that roughening the surfaces of a joint increases the strength of the bond because it provides mechanical interlocking. Current research indicates that the surface roughening increases the bond area for specific adhesion, and that the effect of mechanical interlocking is minimal in many cases. However, the roughened surface is more difficult to wet with the adhesive, and this may result in discontinuities in the adhesive film. Consequently, best results are generally obtained with surfaces that are smooth but not polished.

Most surfaces are contaminated. Even so-called "clean" surfaces are generally coated with a thin layer of foreign matter. Such as an adsorbed film of gas or a thin oxide film. Clean "dry" glass may often contain a very thin water film. These films often interfere with specific adhesion and should be removed in order to obtain the best results with an adhesive. Greasy films can be removed with a solvent wipe or other chemical wash. Xylene or toluene will remove most greasy or oil films. Acetone will also give good results, but is a little difficult to handle at the construction site because it evaporates so rapidly. For metallic surfaces such as steel and aluminum, a light sanding with fine sandpaper will remove the oxide film; then a quick wipe with a clean rag and acetone will remove the sanding dust, and the surface is ready for bonding. For porous substrates such as wood, a fine sanding and a wipe with a clean dry rag will furnish a satisfactory substrate.

Adhesives are generally furnished to the job in liquid or paste form. Some physical or chemical change is necessary to convert the adhesive to a solid. This change may be a hydration, as in a paste of portland cement and water; it may be polymerization, as with the epoxy and polyester resins; or it may be the simple evaporation of volatile materials, as in the solvent-based rubber adhesives. Other methods of curing, such as vulcanization, oxidation, gelation, and pressure reduction, are also used in various applications. Since curing methods vary so widely, some understanding of the properties of adhesives is necessary in order to insure that proper curing methods will be used on the job.

Adhesives are particular. Every adhesive will not form good bonds with every type of substrate. Materials such as wood, the unglazed backs of ceramic tiles, and paper are porous. These substrates have pores or capillaries which will tend to bleed off the vehicle from the adhesive. This can be both an asset and a disadvantage. In some cases, this may result in quick tack and fast drying. On the other hand, a porous wallpaper used with the wrong adhesive could stain very badly. Metal and glass surfaces have little or no capillarity and thus do not confront this problem.

The typical adhesive joint consists of two pieces to be bonded (substrates or adherends), a layer of adhesive, and two interfaces where the adhesive comes into contact with the substrates. In this broad sense, then, a sealant in a joint fits the definition of an adhesively bonded joint. However, the sealant (the adhesive layer) is generally from $\frac{1}{4}$ to $\frac{1}{2}$ in. thick, whereas in adhesive bonding the adhesive layer is quite thin. Research has been conducted on optimum glue line thickness, and the results of these studies indicate that thin glue lines usually perform better than thick ones. The basic criterion for glue lines is that the adhesive must wet the surface; hence any adhesive in excess of the amount needed to wet the surface thoroughly is not only wasteful, but may actually detract from the strength of the joint.

Because of the nature of specific adhesion many materials have a natural affinity toward adhesion. If two thin sheets of glass, such as the slides used with an ordinary microscope, are rubbed together, they will cling together and require some degree of tensile force to separate them. Simple evidence of the effectiveness of thin films can be demonstrated using the same two slides. If a drop of oil or soapy water (which are lubricants) is placed between the same two slides and forced into a thin layer by firmly pushing the slides together, the resulting union may have a tensile adhesive strength of as much as 60 psi. These same adhesive layers would obviously be ineffective if used in thick glue lines.

In any type of bonded joint, some degree of pressure is usually required to keep the parts in intimate contact until the adhesive has cured. This may vary from simple hand pressure to the large presses used in the fabricating of glued-laminated wood members. The proper amount of pressure must be used for any given application. There is no general law which indicates the amount of pressure to be applied; simply, the required pressure must be sufficient to force the surfaces into intimate contact. The surfaces will never bond if the adhesive does not wet both substrates. Bonding pressure depends on the flatness of the surfaces, the type of substrate, the viscosity of the adhesive, and tempera-

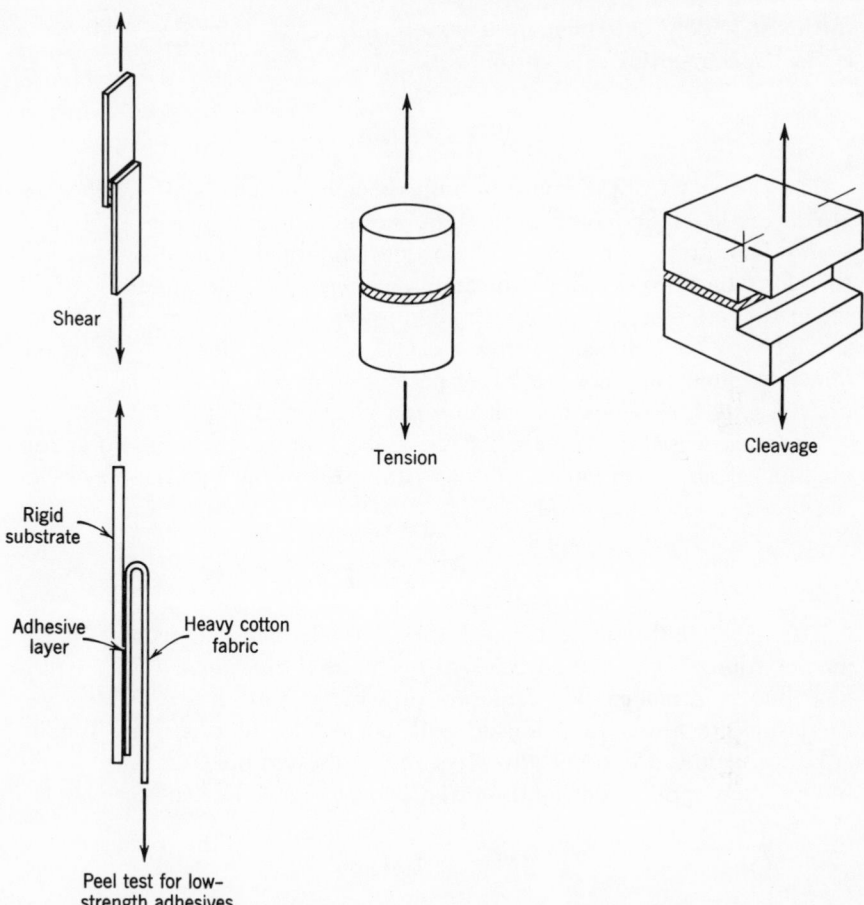

Fig. 21.1 Adhesive test specimens.

ture. There are numerous examples, especially in wood joints, in which insufficient pressure has resulted in a poor bond. However, there are also examples, notably in metal bonding, in which a poor bond has been traced to excessive pressure. In order to obtain the best bond with a particular adhesive, it is best to consult the adhesive manufacturer for his recommendations.

21.7   Testing of Adhesives

Adhesive joints are tested in tension, shear, peel, and cleavage, and sometimes for impact strength and creep. Of these tests, the shear test is the most widely used and most widely respected. Figure 21.1 shows a typical specimen for each of the tests.

21.7.1   Shear

The shear test is a measure of both specific and mechanical adhesion. However, it is conducted on a ½-in. single lap joint which is bonded under laboratory conditions. The values obtained from the test must therefore be considered as maximum values. Job site conditions will lower the values considerably. Bonded joints at the construction site involve larger bond areas and thus a risk of discontinuities in the adhesive film. Also, tolerances for the larger parts used at the job may result in an uneven glue line which is thick in some places and thin in others.

The shear capacity of the adhesive is important in the fabrication and erection of sandwich panels for walls and also in the installation of facing blocks and ceramic tiles.

21.7.2   Tension

The tensile adhesion test may be conducted with sections of 1 in. diameter round bars. The surfaces to be bonded are cleaned and a thin glue line is formed under moderate pressure (10 to 15 psi). After cure, the joints are tested by a tensile pull, normal to the glue line. Tensile adhesion values are a measure of specific adhesion only, and are valuable for such applications as the installation of acoustical ceiling panels.

21.7.3   Cleavage

Values of cleavage strength are usually less important than the other test data for adhesives and are sometimes not reported on the manufacturers' technical data sheets. The loading in the cleavage test is somewhat similar to the load imposed on the 4- × 8-ft sheets of paneling and gypsum wallboard which are installed with adhesives. However, high values of tension and shear are usually accompanied by high cleavage values; hence this test loses a little of its importance.

21.7.4   Peel Strength

The peel strength of adhesives may be measured in several ways. For low and medium strength adhesives (below 1000 psi shear), the same peel strength test used for sealants may be applicable. This test is valid because the adhesive (sealant) is rolled out into a thin glue line. For higher strength adhesives, a small rolling drum may be substituted for the fabric substrate. Peel strength is then measured by rotating the small drum which peels the adhesive from the substrate.

Peel strength is difficult to relate to job site conditions. In the testing of adhesive-bonded joints, it is quite common to see a combination of peel plus tension or peel plus shear as the unit fails. However, a condition of pure peel failure is seldom, if ever, seen on the job. It has been shown, nevertheless, that adhesives with low peel strength generally have low fatigue strength. Also, peel strength is related to notch sensitivity and tear strength. Consequently, discontinuities in the adhesive film are a matter of concern if an adhesive with a low peel strength is being used.

21.7.5   Creep

Just as with sealants, the matter of creep in bonded joints has been treated far too lightly. The adhesive, once it is placed into use, is usually kept under a sustained load. The bonded parts are expected to stay in position for the life of the building. In some applications, such as floor tiles, creep is of negligible importance. However, in structural applications, such as glued-laminated members and metal bonding, creep can be the ultimate design criterion.

The creep test and its companion, the stress rupture test, can yield a great deal of information about the time dependent properties of a bonded unit. The creep test is performed on the same type of specimen used for the shear test. For structural applications, the specimen should be loaded to 50% of its ultimate shear strength. The load is left on the specimen, and the curve of deformation versus time is plotted.

The stress rupture test is somewhat the converse of the creep test. The object of the stress rupture test is to determine the maximum load the specimen will sustain for an indefinitely long period of time. This obviously requires a series of specimens tested under different loads to a time dependent failure. The test should be conducted at a fixed level of temperature. From the results of this test, the manufacturer is able to tell the architect or engineer that a certain adhesive is able to sustain a specified amount of shear loading for an indefinitely long period without failure. This stress rupture test, when used together with the creep test which gives the amount of deformation, provides the architect with an excellent picture of the load carrying capabilities of an adhesive.

21.7.6   Test Limitations

There is much room for development in the testing of adhesives. In the testing of sealants, the differences in specimen shape between laboratory and job site are obvious, and the fallacies in some of the testing methods are therefore quite apparent. In the testing of adhesives, however, most of the tests are at least representative. Two major problems remain. Small size specimens must be tested in the laboratory, for it is seldom practical to test full size units there. It is easy to imagine the problems associated with the testing of a 4- × 8-ft sheet of gypsum wallboard, glued to wooden studs. The other problem can be related to both job site and laboratory. This involves the development of a nondestructive means of evaluating the quality of a bond. Some work has been done using ultrasonic techniques, but a practical nondestructive test which can be used on the job has not yet been developed.

Temperature has been mentioned very little because it affects different adhesives in different ways. Thermoplastic adhesives, such as the bituminous roofing cements, soften with heat and become hard and brittle in cold weather. Thermosetting adhesives, such as the epoxies and polyesters, will tend to embrittle at cold temperatures. At higher temperatures these materials do not soften and melt, but they do lose some strength. Actual values will be stated in the subsequent chapters about the various materials.

21.8 Specifications

At least 50 different specifications for adhesives are applicable to the construction industry. In addition to specifications, there are standards, such as those of the American Institute of Timber Construction, which carry as much weight as Federal specifications within a given trade. There are Federal and Military specifications, ASTM specifications, USASI specifications, and there are the standards of various groups, such as the Timber Institute, the Tile Council of America, and many others. Consequently, even the most experienced architectural specification writer finds it extremely difficult to have even a bare working knowledge of the requirements of all the adhesive specifications.

Fortunately, in the construction industry there are virtually no adhesive specifications which cross building trades. Requirements for tile setting adhesives are determined by one group and adhesives for wood by a separate group. The Federal and Military specifications are used as guides by the specification groups in the various trades, but rarely, if ever, does the architect specify an adhesive. Rather, the architect would specify the adhesive recommended by the tile manufacturer, in order to place responsibility for the finished job on the manufacturer or the installer of the tile. The architect, in specifying the use of glued-laminated (glulam) timber members, would refer to the standards of the American Institute of Timber Construction. The laminator of the structural members then assumes responsibility and uses adhesives which meet AITC standards.

The exception, the case in which the adhesive is usually specified, is in the bonding of cementitious materials. The architect specifies the bonding agent to be used in the installation of stone, quarry tile, and cement floors. The highway or bridge engineer specifies the adhesive to be used for the repair and rehabilitation of damaged concrete and other structural bonding. The adhesive most often specified is an epoxy. The specifying authority may be an architect, a consulting engineer, or the materials engineer of a state highway department. There are Federal and Military specifications which cover the epoxies. Consequently, the finished specification will usually be a performance-type specification, either based on the Federal specification, or else written around the properties of a particular product which has proved itself in test installations.

The tile adhesives are often hallmarked by the Tile Council of America. The TCA maintains a research institute which assumes much of the responsibility for the quality of the adhesives. The hallmarks are related to existing standards, such as Department of Commerce CS 181–52 (USAS1 181–52). The main features of these standards are a shear test and a test for the durability of the bond. The shear test is conducted under four sets of conditions:

1. 16 hours after bonding
2. 7 days after bonding
3. 7 days air drying followed by 7 days of water immersion
4. 28 days after bonding (40 psi shear)

The durability test for the quality of the adhesive bond is a cyclic weathering test modeled after ASTM-D-1037.

The lumber industry depends on specifications which are based mostly on research and testing done by the Forest Products Laboratory of the Department of Agriculture. Groups such as the Douglas Fir Plywood Association and the American Institute of Timber Construction base their standards on Federal and Military specifications which were developed by FPL. Ninety percent of wood gluing applications fall under two headings: casein glue, which is used for interior work with low humidity environments, and resorcinol glue, which is used for exterior work or high humidity applications. Casein glues are covered by Federal Specification MMM-A-125. Resorcinol glues are covered by Federal Specification MMM-A-181 or Military Specification MIL-A-5534A. These wood specifications are, at present, captive specifications.

There is little likelihood that the architect will ever specify the type of adhesive used to hold plywood together. However, a great deal of research is being conducted into glued-nailed trusses and glued subfloorings and therefore the architect may have to familiarize himself with these specifications in the near future.

In the field of metal bonding, the aircraft industry has made use of the following Military specifications for structural metal-to-metal airframe adhesives and for the metal-to-metal bonding of structural sandwich materials.

MIL-A-5090: adhesive, airframe, structural, metal-to-metal
MIL-A-25463: adhesive, metal-to-metal, structural sandwich material

These specifications appear to be quite adaptable to the needs of the construction industry. Metal bonding is beginning to show up in building construction, and the architect will in the future become responsible for specifying the proper adhesive.

22

Flooring Adhesives

Adhesives for flooring materials are high volume, low cost materials. These adhesives are generally manufactured by the same companies that manufacture the flooring materials. The adhesives are advertised and promoted as a unit with the flooring and are generally so specified by the architect. The type of adhesive, of course, varies with the type of flooring and also with the type of substrate (subflooring) over which it is to be installed.

There are nine types of flooring material which require adhesives for their installation. Table 22.1 shows the estimated 1967 sales of these various types of flooring materials.

In 1966, almost a million pounds of adhesives were used for the installation of flooring materials: 60% of the flooring (and adhesive) sales were in the replacement market, and roughly 40% were used in new construction. The flooring materials use various adhesives, but linoleum

Table 22.1

Type of Tile	Millions of Square Feet
Vinyl asbestos	1,380
Inlaid vinyl sheet goods	710
Asphalt	525
Sheet linoleum	300
Wood block (including mosaic parquet)	275
Vinyl	160
Rubber	34
Linoleum	26
Cork	5

paste, rubber, asphalt, and polyvinyl acetate are the most commonly used classes of adhesives. Specific materials vary widely and include solvent-based asphalt mastics, hot asphalt-based materials, asphalt emulsions, sulfite liquor adhesives, polyvinyl acetate (white glue), rubber-based mastics, resin-based mastics, neoprene contact cements, and rubber emulsions.

The substrate for a flooring material is generally either a wood subfloor or a concrete slab. Wood subfloors are seldom used in below grade applications, but concrete slabs may be used either above or below grade. Consequently, adhesion to wood and concrete and sensitivity to moisture are the prime considerations in the selection of a flooring adhesive. When concrete slabs are in contact with the ground, the slabs tend to remain moist. The moisture which is present in the slab is alkaline in character. This alkalinity of the slab surface may have an adverse effect on adhesion.

Even though flooring materials and adhesives vary widely, a few general principles should be observed. For installations on wood subfloors above grade, there is some latitude in the choice of adhesives. Asphalt tile and parquetry wood blocks can successfully be installed with the asphalt-based mastic adhesives. However, the asphalt-based adhesives should not be used with rubber or vinyl products. Rubber-based mastics, resin-based mastics, and white glues can be used to install both the sheet goods and the floor tile units. Contact cements such as the neoprenes are also very successful in the installation of all types of floor tiles, including the parquet blocks. The neoprene contact cements are also the best choice for the installation of molded vinyl and rubber stair treads and mats. Sheet flooring, such as linoleum or inlaid vinyl, is best laid with a mastic-type adhesive. The contact cements are difficult to work with when handling large sheets of flooring material. Figure 22.1 shows the installation of a sheet vinyl flooring, using a mastic adhesive.

For installations on wood subfloors, the mastic adhesives are generally spread with a notched trowel. A trowel with notches $1/16$ to $1/8$ in. deep spaced on $3/8$-in. centers will serve adequately. A clean subfloor is required, but no priming of the surface is necessary. A lining felt is sometimes placed over the subfloor before installation of the flooring material. The contact adhesives may be installed by notched trowel, but are also available in a brushable consistency. In some cases, it may be necessary to apply a thin brush coat of the adhesive to seal the subfloor and prime the surface. Wood parquet blocks are now available with the contact adhesive already on the wood block. If the underlayment of hardboard or plywood is sound and true, the release paper can be peeled from the back of the blocks and the blocks can be set immediately. If the underlayment is rough, it should be brush primed, whereby much of

Fig. 22.1 Installation of a sheet vinyl flooring using a mastic adhesive. (*a*) Spreading the mastic. (*b*) Rolling the flooring. (*c*) Finishing the seams. (Courtesy of Armstrong Cork Company)

(c)

the advantage (and expense) of this type of flooring is lost.

In residential construction much of the resilient flooring is installed on concrete subfloors, which may be slabs resting on the ground or basement slabs which are below grade level. In commercial and industrial construction, the majority of applications on concrete slabs are above grade level. In all applications of flooring to a concrete substrate, the major factors affecting adhesion are moisture and alkalinity. However, these problems are much more severe in the on-grade and below-grade applications. According to Rohrer [16], one or more of the following situations may be encountered:

"(1) The adhesive film may be dissolved or chemically attacked by the alkaline moisture.

"(2) If a subfloor is wet at the time the installation is made, the adhesive will not dry or set properly. In this case an adequate bond is never obtained.

"(3) The dry adhesive film, even though resistant to alkaline moisture, may be stripped from the sub-floor by water. . . . The result is that the adhesive film is really floating loose on a layer of water.

"(4) The adhesive film may be softened and stripped from the sub-floor by moisture. Subsequent foot traffic will then force the adhesive through the seams or joints of the floor covering. This effect is more likely to occur with the asphalt adhesive used to install asphalt tile and vinyl asbestos tile."

Some architects have used membrane waterproofing under slabs on grade to prevent the entry of moisture into the slab. Heavy seal coats have also been used on top of the slab to provide a moisture barrier. For concrete slabs on or below grade, water sensitive adhesives, such as polyvinyl acetate, should never be used. Research has shown that the rubber-based mastics with a water vehicle (rubber latex) will give good performance in these critical applications. For concrete slabs above grade, if the slab is permitted to dry thoroughly, more latitude is possible in the choice of adhesives. Depending on the type of floor covering being used, resin-based mastics, rubber-based mastics, asphalt-based adhesives, and polyvinyl acetate may all be used.

Wood block flooring (end grain blocks) for industrial floors presents a specialized application. For large installations (over 2000 sq ft) these blocks are usually set in a hot coal tar pitch. The pitch is heated to 400 to 425 F, and spread by squeegee. This pitch has a very short "open time" (less than 1 min), so the melted pitch must be spread rapidly. After the pitch has set, the blocks are placed and the surface is flushed over with a filler, applied by squeegee to fill all the joints in the floor. For small installations and occasional replacement of block, a viscous asphalt emulsion can be trowled onto the subfloor. After the adhesive has dried sufficiently to become quite tacky (about 1 hr), the block may be set.

Any flooring installation is no better than the surface to which it is applied. Concrete floors should be as dry as possible. Grease and oil stains should be removed. Any loose dirt and cement dust should be removed by sweeping or by vacuum-cleaning, and any protrusions on the concrete slab should be ground off to the level of the slab.

Flooring adhesives are typically low or moderate strength adhesives (less than 1000 psi shear). The loading on flooring adhesives is compression with little or no shear. The adhesive may be subjected to some flexing when installed over wood subfloors. Virtually all installations are interior use; consequently, thermal expansion and contraction are not too much of a problem.

23

Wall Covering Adhesives

23.1 Ceramic and Plastic Tile

Ceramics in various forms have been used by construction men since ancient times. In this country glazed, or ceramic, tile units have been used as a wall covering for almost 100 years. The tile units are made from finely ground and prepared clay which is bisque-fired to a hardened but somewhat porous state. The faces of the tile units are coated and fired again (vitrified) so that one surface has a hard glassy finish. The backs of the tile units remain unglazed and somewhat porous.

In the early days of construction the tile-faced wall was considered in the same category as any other faced masonry wall. Consequently, the tiles were set in the same manner that a brick or stone facing was set, namely, in a bed of mortar. Sometimes ordinary lime mortar was used, but it soon became standard practice to set the tiles in a portland cement mortar. Since the backs of the tiles were porous, some slight advantage was obtained from mechanical adhesion. Portland cement tile installations were, and still are, a quite satisfactory method of setting ceramic tile. Within the last 20 years adhesives have been widely adopted for tile setting. The adhesives are faster than the mortar method, and they place less dead load on the wall structure. In residential construction, adhesives are better suited for the installation of ceramic tile over gypsum wallboard.

Ceramic tiles offer many unique advantages as a wall covering and consequently, the market is still growing. In 1965 approximately 200 million sq ft of ceramic tile were used by the construction industry. The adhesives used for the installation of these tiles are either solvent-based rubber mastics or latex emulsion systems. The adhesives for wall tiles are not manufactured by the same companies that manufacture the tiles. However, tile manufacturers will recommend specific adhesives and architects will generally specify in accordance with these recommendations.

The wall tile adhesives are typically low to moderate strength, low cost adhesives. The latex emulsion systems, such as PVA, sell for about $2.50 to $4.00 per gallon. The solvent-based mastics are slightly more expensive.

The wall tile adhesives should have a fast grab, but since the adhesive must support only the dead load of the tile, high shear strength is not required. USASI Standard 181–52 requires only 40 psi in shear. The adhesive should have enough resistance to creep so that the tile will not slip out of alignment. However, the shear load of the tile on the bond line is not excessive, so that both the solvent and emulsion adhesives meet this requirement readily.

There is one specification requirement for the wall tile adhesives which is quite demanding. Standard 181–52 requires a shear test of the adhesive after seven days of drying and seven days of water immersion. This is an excellent specification requirement, because ceramic tile is often used for bathrooms, locker rooms, showers, and other installations in which it is exposed to high humidity conditions. The solvent-based adhesives usually meet this requirement, but very few of the latex emulsions are able to pass this test.

In extremely critical water immersion installations, such as swimming pools, a two-component neoprene-based adhesive may be specified. This type of adhesive has a higher shear strength than the other ceramic tile adhesives, as well as better water immersion resistance. This adhesive may also be a good choice for exterior ceramic tile applications. Figure 23.1 shows a shipboard installation of ceramic tile using the neoprene adhesive.

Since World War II plastic tiles have come into use in residential construction. The plastic tiles are light in weight, and easy to handle and install. However, the plastic tiles are not as durable as the ceramic tiles.

A different set of standards covers the adhesives for plastic tile, and a different hallmark is issued by the Tile Council of America. The water immersion resistance required of the plastic tile adhesives is the same as for the ceramic tiles. The load-bearing capacity required of the plastic tile adhesives is less than for the ceramic tiles, because the light weight plastic tiles do not exert as much dead load stress on the bond line.

Fig. 23.1 Shipboard installation of ceramic tile using a two-component neprene-based adhesive. (Courtesy of Crossfield Products Corporation)

23.2 *Wallboard or Panel Adhesives*

Wallboards and paneling, in the majority of cases, are still installed by nailing to studs or furring strips. As better adhesives have developed, adhesive installations have taken over a slowly increasing segment of the market. Adhesives do offer definite advantages in many cases. Paneling may have an expensive veneered surface of wood or a decorative laminate in which nail holes are almost impossible to disguise. Gypsum wallboard, when nailed into place, requires that the nail heads be in-set ánd then plastered over and sanded before painting.

Adhesives are currently used for the installation of gypsum wallboard, plywood (including paneling), hardboard, vinyl-clad steel laminates, asbestos cement board, and low density insulating fiberboard. The substrates to which these boards are attached may vary, but wood is by far the most common. The adhesives used, then, must have good adhesion to wood as well as to the type of wallboard being installed. Such adhesives include mastics, contact cements, and white glue (PVA). If the building wall or furring line is fairly smooth and true, the contact adhesives may be used. If the furring strips are ragged, or if a rough textured wall such as concrete block is being covered, an adhesive such as the mastics or white glues, which have a gap filling capability, must be used.

The mastics are high solids adhesives with a fast grab. They may be based on asphalt or of a reclaimed rubber resin blend. The vehicle for the rubber adhesives is a very volatile solvent and thus good ventilation is required during installation. The rubber mastics are more expensive than the asphalt-based materials, but their lighter color offers an advantage in terms of both spotting the finished surface and operator's clean-up. With regard to strength, either type will do an adequate job. The mastics may be installed by caulking gun, knife, or trowel, but the caulking gun is the most popular method among the tradesmen. Figure 23.2 shows a typical caulking gun installation. Figure 23.3 shows a special long barrel caulking gun, favored by drywall applicators for ceiling and wall work without scaffolds.

The advantages of the mastics include the following: fast grab; enough body to compensate for surface irregularities; since they are a relatively thick glue line application, they need be applied to only one of the mating surfaces; the panel can be moved slightly for proper positioning after the panel is placed; they are relatively inexpensive.

Fig. 23.2 Application of a drywall adhesive with air-operated caulking gun. (Courtesy of Pyles Industries, Inc.)

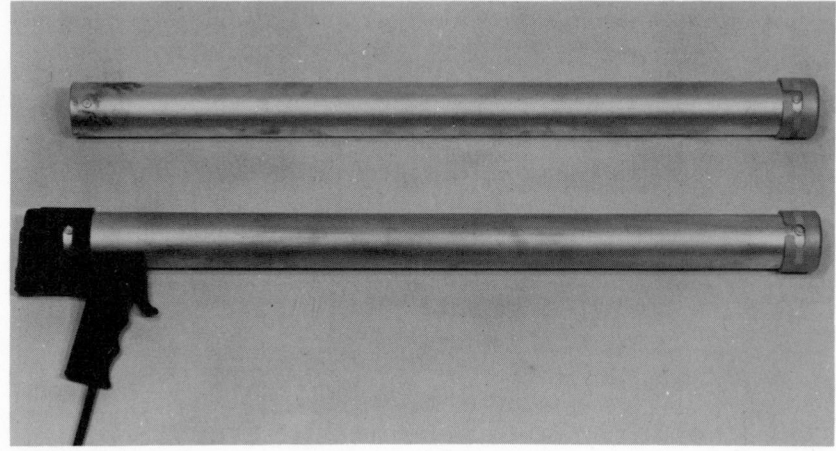

Fig. 23.3 Long barrel caulking gun for adhesive installation on ceilings. (Courtesy of Pyles Industries, Inc.)

However, the mastics may often require some bracing to hold them in position until the glue comes up to strength, and they may also require some supplemental edge nailing. The mastics are not well suited for thin film applications, such as the installation of semirigid decorative laminates.

The contact cements, on the other hand, are intended for thin glue line applications. They are generally viscous liquids with a neoprene base and are applied by brush or spray. The contact cements contain a high percentage of solvent for easy spreading. After the adhesive has been applied, the solvent must be allowed to evaporate before the panel is placed. The surface of the contact adhesive may be tested by lightly touching the surface with a finger. If the adhesive has lost its tacky "feel," the panel may be positioned and pressed into place with moderate hand pressure or a clean roller.

The advantages and disadvantages of the contact cements are almost the converse of the mastics. The contact cements provide an instant bond with no supplemental bracing or edge nailing, as well as a good bond with a thin glue line. However, it is not possible to move the panel once it is placed, so that alignment of large sheets may be quite difficult. Because of the thin glue line, the adhesive is usually applied to both of the mating surfaces. Also, the contact cements should be used only where the substrate is sound and true to line, because these glues have no appreciable thickness for filling gaps.

The white glues are emulsions of polyvinyl acetate which are easily applied by knife, trowel, or caulking gun. These adhesives have enough strength to do an adequate job, but do not have the fast grab of the other adhesives and also are not suitable for high humidity applications. These adhesives offer the advantages of easy application and fast clean-up. Hands, tools, and spills can be cleaned quickly with warm water. However, these glues take longer to set and, consequently, require bracing of the panels until the glue comes up to strength. The white glues can be used for thick film applications if the mating surfaces do not conform perfectly.

23.3   *Other Wall Coverings*

Decorative semirigid sheets of plastic laminates, such as Formica °, have been used for many years as coverings for counter tops and work surfaces. In recent years, however, the uses of these laminates have been expanded. Some of the plywood paneling available on the market has a plastic laminate surface instead of wood veneer. Kitchen cabinets made from flake board or particle board coated with plastic laminates are now available. Sheets of the plastic laminate approximately $\frac{1}{16}$ in. thick are now used as wall coverings. Entire kitchens can now be ordered in which the kitchen cabinets and the wall covering use the same laminate with a wood grain finish. The same plastic laminates are finding increasing use as a wall covering in laboratories, classrooms, corridors, and other hard usage areas.

These semirigid laminates can be installed successfully only by using adhesives. The adhesives usually used are the contact adhesives based on neoprene. The laminates are supple enough to bond smoothly over the average plaster or gypsum board wall surface. The laminates are delivered to the job in 4- × 8-ft sheets, and are installed in relatively large pieces. Once the sheet is in place, the dead load stress on the bond line is negligible. There is no current specification for the contact adhesives used with the plastic laminates.

In the field of flexible wall coverings, wallpaper held in place by the familiar starch-based wallpaper paste remains the standard. Over 10 million lb of wallpaper paste were sold in 1966, and the greatest percentage of this total was sold to professional paperhangers. The average homeowner would not hesitate to paint his own living room, but would call in a professional if the walls were to be papered. The versatility of wallpaper has greatly expanded in the past 10 years. Papers with a washable plastic finish which are installed with wallpaper paste are cur-

° Trademark of Formica Corporation.

rently available. Other materials which fit roughly in the wallpaper category are fabrics such as burlap and plastic-coated fabrics. Modern decorators are now even using carpeting as a wall covering for playrooms and recreation areas. These wall coverings are generally installed with a rubber-based adhesive with a fast grab. Contact cements are not recommended because the wall coverings are flexible and the pieces must be shifted slightly on the wall in order to match patterns.

The adhesive and wall covering industries have devoted considerable effort to capturing the "do-it yourself" segment of the wall covering market. Precoated wallpapers of various types were tried and found unsuccessful. In recent years, vinyl- and plastic-coated paper wall coverings, precoated with contact cement, have met with some success. These contact wall coverings are packed in rolls with a protective release liner. These coverings are tough and washable, and homeowners are using them for covering counter tops, furniture, and even books, as well as walls. Finishes are available in many colors, wood simulations, and imitations of brick and stone.

Compared to ordinary wallpaper, these coatings are expensive. In addition, because they bond with contact cements, positioning is critical. The piece cannot be moved once it is placed. Because of these limitations, the use of this adhesive product is generally limited to small areas.

24

Roofing Cements

The roofing industry accounts for the largest volume of adhesive products used by the construction industry today (about 370 million lb in 1966). The line of distinction between sealants and adhesives can be very fine, and nowhere is this more apparent than in the roofing industry. The roofing cements are mainly asphalt- or coal tar-based materials. The uses of the roofing cements span the entire construction industry, including both residential and commercial construction.

In residential work, the primary roofing material is usually shingles, nailed into position. However, roofing cements are used for the cementing and sealing of vent stacks, valley and chimney flashing, eaves, and gutters. The roofing cements used for residential work are usually solvent-based, asbestos-filled materials which are cold applied by knife or trowel.

The shingles used in residential work may be slate, wood or asbestos cement which are rigid, or asphalt shingles which are quite flexible. The asphalt shingles are the most widely used because of their ease of installation and low cost. Asphalt shingles which contain a tab of adhesive on the lower surface of the shingle butt are currently available. This adhesive holds down the leading edge of the shingle to prevent the shingle from being blown back and folded over in very high winds.

In commercial and industrial construction, the hot applied roofing cements account for the greater volume. The cold applied solvent-based materials may be used for the cementing and sealing of cant strips, roof drains, vent stacks, and flashing, but the hot applied materials are used for cementing the bulk of the roofing material into place. For example, a building 100 × 100 ft contains 10,000 sq ft of roof surface, or 100 "squares" of roofing. A five ply built-up roof for this building contains six layers of adhesive, one for each layer of roofing felt and one for the aggregate surfacing. Depending on waste and layer thickness, this building might use 1000 gal of hot roofing cement.

The hot applied roofing adhesives should be heated in a jacketed heater. A temperature of 400 to 425 F is sufficient to make the material fluid enough for easy application. Overheating is not as critical as with the hot poured highway sealants which contain substantial amounts of rubber; nevertheless, the roofing cements should not be overheated. Temperatures above 450 F will degrade the roofing cement. The hot cements are generally applied by mop. The roof deck is first mopped with a layer of cement. The roofing felt, supplied in rolls, is placed in strips and the laps are mopped to insure a watertight roof. The final coating of roofing cement may be somewhat thicker than the other layers in order to give it a good "bite" on the aggregate surfacing.

There are other materials which are used minimally in roof construction. Sealants are sometimes used as adhesives for the bedding of skylights, and so forth, but in volume applications no other material competes with the coal tar and asphalt cements on a price per performance basis.

25

Sealants as Adhesives

The adhesively bonded joint consists of two substrates with a layer of adhesive between them. The sealed joint consists of two substrates enclosing a bead of sealant. The apparent difference between the sealant and the adhesive, then, is partly in the function to be served and partly in the thickness of the bonding layer.

Many sealants are versatile enough to be used for both applications. One of the more versatile materials in this category is the latex caulk based on PVA. This material serves well as a sealant in joints with moderate amounts of movement. The same material might be classed as one of the "white glues." This material when tested in the laboratory as an adhesive shows shear strength values higher than those required by Standard 181–52 for tile adhesives. Homeowners are now using this material not only for caulking and sealing, but also for sticking a loose wall tile back into place, as well as other adhesive uses.

Elastomeric sealants can also be used as adhesives. Silicones, polysulfides, and urethanes can be used for setting and bedding skylights, bonding name plates to office doors, setting screw anchors into masonry walls, forming cast-in-place door bumpers, and other miscellaneous adhesive uses around construction jobs. The familiar silicone bathtub sealant is finding increased usage around the home as an adhesive. This material can even be used for the bonding of patches onto children's play clothes [17].

The uses of sealants as adhesives on the construction job are not volume applications. They are mostly odds and ends which must be taken care of on the spot. The object here is simply to point out to the construction man that, if he has two things to stick together and has a good quality sealant on the job, there is usually no need to run back to the supplier for a cartridge or two of special adhesive. The chances are pretty good that the sealant will do an adequate job.

26

Pressure Sensitive Adhesive Tapes

There is hardly a home or office in the United States which could function efficiently without a roll of pressure sensitive adhesive tape. The familiar transparent tape is used for sealing packages, tacking up notices, repairing torn books, and hundreds of other uses. Ingenious contractors have begun to realize the versatility of the tapes and are using them more every day. The adhesive manufacturers have kept pace with this progress and now produce dozens of tape types to meet the contractor's needs.

Pressure sensitive tapes serve as both sealants and adhesives, depending on the application. The tapes can be used to effect a permanent seal, but can also offer easy removability for temporary construction applications, such as the sealing of concrete form liners.

These tapes consist of a backing material coated with a pressure sensitive adhesive, so that the tape can be placed with only a moderate hand pressure. Backing materials range from paper and cloth to aluminum foil and films of polyester and other plastics. Backings reinforced with yarn are common, as are combination backings such as metal foil and paper laminates. The thickness of the tape backing varies from about three mils (0.003 in.) up to about 15 mils (0.015 in.). Some resin-based adhesives are used, but most of the tapes use a rubber-based adhesive.

The tapes vary in strength from about 20 to 500 lb per inch of tape width. Tapes with reinforced backings usually have the highest values of tensile strength. This tensile strength, however, is the strength of the backing material and not the adhesive strength of the bonding layer. There are a few tapes which use high strength thermosetting adhesives, but the tapes which are used most often in construction are character-ized by low strength adhesives. These tapes yield values of 10 lb or less per inch of width when tested in shear. For permanent applications, the strength of the backing and the strength of the adhesive layer should match fairly well. For temporary construction usage, it is generally ad-visable to use a tape with good backing strength but low adhesive strength, to provide easy removability. Consider, as an example, the re-painting of an older building. The painter will use masking tape to mark off the borders of different colors. The tape should have enough backing strength to stand abuse, so that it can be put under some tension and placed in a straight line. Obviously, the tape must be removable. If the adhesive strength of the tape is too high, the tape may well pull loose some of the previous coat of paint when the tape is removed. The painter then has an expensive patching and touch-up job on his hands.

Tapes which are used for permanent applications should have good aging and weathering characteristics. Tapes for both permanent and temporary construction use usually require a high degree of resistance to water. Some tapes, especially those used in duct work, should also have good abrasion resistance. Table 26.1 is a summary of most of the tapes currently available for construction use.

26.1 Testing and Specifications

The tapes for construction use are a relatively new product in terms of volume usage. Consequently, these products are being intensively tested by the manufacturers before marketing. The usual tests conducted on the adhesive tapes are as follows:

1. Tensile strength of the backing material
2. Shear strength of the adhesive
3. Peel strength of the adhesive
4. Aging and weathering
5. Creep of the adhesive under constant load

The manufacturers have developed their own performance require-ments for the tapes. Since there are so many types of tapes for different applications, there are actually no standard specifications the architect or contractor can depend on for help in specifying. Consequently, the great majority of architects and contractors tend to specify and buy tapes by brand name. This buying tendency has channeled the tape market to a relatively few large manufacturers who produce a wide range of products.

Backing Material	Adhesive	Tensile Strength (lb per in. width)	Elongation (%)	Total Thickness (mils)	Adhesion to Steel (oz per in. width)	Temperature Range (deg F)	Remarks
Paper							
Creped Kraft, impregnated	White rubber resin	21	7	6.5	34	32 to 150	Commonly called "masking tape." Inexpensive. Long shelf life. Can withstand 350 F for 30 min.
Rope fiber, flat-paper, color impregnated	Rubber resin	40 to 62	3 to 4	6.5	35 to 50	32 to 150	High strength combined with high adhesion. Available in colors and can be used to identify as well as seal. Used in packaging.
Impregnated creped paper, reinforced with glass yarn	Rubber-based	270	4	13.7	45	32 to 150	High tensile and medium impact strength.
Cloth							
Cotton	Rubber-based	40	6.5	12 to 14	30	32 to 150	Good tensile, tear, and abrasion characteristics. Often can be reused. Excellent shelf life. Good moisture resistance. Conforms well to irregular surfaces.
Vinyl coated cotton	Rubber-based	65	...	14	32	...	Excellent aging characteristics; good resistance to oil, heat, and cold; fair solvent resistance. Good moisture resistance. Good for exterior sealing.
Polyethylene coated cotton sheeting	Rubber resin	60	7.5	15	50	...	High-strength, high-hold waterproof tape.
Glass cloth	Thermosetting rubber-resin	150	5	7	40	260 continuously 320 for 2 wk 390 for 2 hr	Requires heat cure: 3 hr at 250 F, 2 hr at 275 F, 1 hr at 300 F. High strength at high temperature. Resists solvents such as toluol and xylol.
Glass cloth	Rubber resin	120	4	5	30	250 continuously 400 short periods	Good tear resistance, tensile strength, and flame resistance.

Table 26.1 (continued)

Backing Material	Adhesive	Tensile Strength (lb per in. width)	Elongation (%)	Total Thickness (mils)	Adhesion to Steel (oz per in. width)	Temperature Range (deg F)	Remarks
Glass cloth	Thermosetting silicone	150	5	6.5	15 @ 73 F 25 @ 73 F for 72 hr 45 @ 250 F for 72 hr 45 @ 500 F for ½ hr	−65 to 450	Adhesive requires cure at room or elevated temperatures. Good solvent resistance with room temperature cure. Max solvent resistance with 24 hr cure at 500 F. Long shelf life. High strength and good holding at high temperature.
Metal Aluminum foil	Synthetic	30	5	5	50	−65 to 250	Reflects both heat and light. Low vapor transmission rate (0.1 gr of water per 100 sq in. per 24 hr). Good aging characteristics. High coefficient of expansion.
Aluminum foil	Synthetic	50 80 120	13 13 13	8 13 17	65 65 65	−65 to 250	Excellent weathering characteristics. High resistance to chemical attack. For permanent applications. Good for vibration dampening.
Cellophane Tape	Rubber resin	30	12	2.7	30	Room	Common transparent mending tape. Least expensive.
Cellulose Acetate Clear film	Clear mass	25	12	2.8	30	. . .	Resists water. Excellent dimensional stability; good weathering ability. Backing is more brittle than cellophane.
Acetate film (0.38 mil) laminated to impregnated rope fiber	Special moisture-proof	44	4	7	35	. . .	Has high hold adhesive. Moisture resistant: excellent dimensional stability.

238

Material	Adhesive						Characteristics
Acetate film (1.5 mil) reinforced with rayon strands	Rubber-based can be colored	210	12	11.5	45	...	High impact and high tensile strength. Very difficult to break. Water resistant.
Acetate film (1.5 mil) reinforced with fibre glass filaments	Rubber-based,	375	4	9.5	50	...	Good holding to glass, wood, steel, aluminum, and polyethylene.
Polyester Transparent film	Synthetic elastomer	25	110	2	24 to 40	−60 to 325	Excellent chemical and thermal stability. Resists many solvents.
Glass yarn re-inforced film	Rubber resin	425 to 500	3 to 4	7.5 to 10.8	40 to 45	−32 to 150	For outdoor weathering conditions and abrasion resistance. Black color protects against ultraviolet rays. Maintains high tensile and holding strength for long periods. Waterproof.
Metallized film	Bonding synthetic	25	110	2.5	40	−60 to 300	Excellent chemical and thermal stability. Metallic coating is aluminum deposited between film and adhesive. Excellent color stability (2 years or longer in outdoor conditions).
Film reinforced with rayon	Rubber-based	225	13	10.8	45	...	Excellent holding properties on glass, wood, steel, aluminum, and polyethylene.
Vinyl Pigmented plastic	Rubber resin	18 to 20	150	6 to 7.5	25 to 25	...	Withstands water immersion; flame resistant; abrasion resistant. Resists grease, oil, and many solvents.
Silvery-gray plastic	Synthetic rubber	20	135	6	21 longitudinal 33 transverse	...	Flame resistant. Excellent for warm or cold air ducts; reduces noise transmission. Good vapor barrier. Seals irregular surfaces tightly.

26.2   Applications of Adhesive Tapes

The adhesive tapes are used for both permanent and temporary construction. The applications of the tapes are so diverse that it is impossible to list them all, but enough applications will be listed here to show what the tapes can do.

One permanent application of tapes is in the sealing of lap joints in metal roof decking. The tape for this application may be a metal foil tape to match the properties of the deck being sealed. The metal roof deck often serves as a base for a hot applied built-up roof, so the tape adhesive must therefore have high heat resistance.

Another permanent application of adhesive tape is the coating of pipes which are to be laid underground. The tape is easily applied and provides permanent corrosion resistance for the pipe. Since the pipe is to be buried in soil which may be wet, both the tape backing and the adhesive must have good water immersion resistance.

Tapes are also used to hold preformed insulation in place around piping. This application is often found in residential construction. The pipes are coated with insulation to prevent condensation. If the pipes are to remain exposed, as in a typical basement, the tape backing should be paintable so that the basement will have a finished appearance.

Heating and air-conditioning ducts are quite often joined with pressure sensitive tapes. The weight of the duct work is supported by the building frame, but the tape serves as both adhesive and sealant to join the sections of duct work.

Residential builders sometimes use 2-in.-wide tapes to provide draft-free floors. These tapes are used to seal the floor-to-wall joint and then are covered by the base molding.

Some tapes have become quite commonplace on the construction job. It is probably safe to say that no electrician would walk onto a job without a roll of adhesive tape in his tool kit.

Temporary uses of tapes in construction probably outnumber the permanent applications. The "closing-in" of a building during cold weather is often accomplished by large polyethylene sheets. These sheets are very conveniently jointed by 2-in.-wide strips of polyethylene tape.

Often, additions to existing buildings must be made with a minimum of interference to the operation of the existing building. In this case, temporary partitions must be erected to keep construction dust out of the existing building. The most convenient way to seal these partitions is by using pressure sensitive tapes.

Tapes are also often used to seal concrete formwork. Sometimes wooden forms are lined with polyethylene film. These sheets of film can be joined with adhesive tapes. The contractor may also choose to apply tape strips directly to the seams in the wood formwork.

Masking tape is used by painters and caulking contractors to form neat lines and joints.

Tapes are also used to seal and protect units, such as louvred windows, during shipment and storage at the job site. Tapes are used to protect expensive finish hardware until it is placed in the building.

26.3   *The Future of Adhesive Tapes*

During the past five years, the use of adhesive tapes in construction has grown almost four times as fast as the overall construction industry. Contractors now realize the convenience value of the tapes and are using them more and more. Temporary uses of tapes should continue to grow rapidly. However, tapes for permanent applications will probably show slower growth. Any component which remains as a permanent part of the structure will be subject to regulation by specifications and local building codes. This regulation will lead to standardization of manufacture and testing for certain types of tapes, which will somewhat slow down the growth. In the overall picture, the tapes should continue to grow at least twice as fast as the construction industry for at least the next 10 years.

27

Adhesives as Binders for Composite Materials

Composite materials are those which combine two or more dissimilar materials so that the system acts as a unit. There are dozens of examples of composite materials actively being used by the construction industry. Concrete consists of sand and stone glued together so that it acts as a unit. Particle board and hardboard are made from small pieces of wood held together by an adhesive. Composite beams are formed by casting a concrete slab on top of a steel beam; in this case, the two elements are held together by a shear connector so that they act as a unit. Structural sandwich panels are also composites.

Composite materials have many advantages, and the use of these materials is growing. Every material known has its strong and weak points, and composites often combine materials in such a way that the best use is made of each element. Concrete is good in compression, but weak in tension. Consequently, the composite beam places the concrete slab into compressive loading, and the steel beam section takes the tension. Composites often make use of very low cost raw materials and combine them into high quality finished products. Particle board uses wood chips which are actually mill waste from the lumber industry. These chips act as a fibrous reinforcement for low cost phenolic, or urea resin adhesives, and are formed into sheets for counter tops, kitchen cabinets, and other construction uses in which sheet materials are required. These hardboards can be sawed, shaped, and nailed just like lumber. Approximately 350 million lb of adhesives were used by this industry in 1966. This is a captive use of adhesive, however. The adhesive used is the choice of the manufacturer. The architect or engineer never specifies the adhesive to be used in a hardboard.

242

Thermal insulation might also be considered a composite material. Adhesives are used to secure paper backings and vapor barriers for batt-type insulation. Other types of insulation also use an adhesive to hold the fibers together. This is another high volume use of adhesives in construction (115 million lb in 1966); but this, also, is a captive use. The architect or engineer checks the heat loss rating of the insulation and specifies the type and amount of insulation to be used, but never specifies the type of adhesive to be used in making the insulation.

27.1 Polymer Concrete

Concrete is an artificial conglomerate rock. It consists of sand and stone glued together. If the aggregates are glued together with a paste of portland cement and water, the result is portland cement concrete. If the aggregates are glued together with asphalt, the end product is asphaltic concrete or "black top." If the aggregates are held together by a polymeric binder, such as an epoxy or a polyester, the result is polymer concrete. The polymer concretes are just beginning to receive widespread use in construction. Approximately two million lb of epoxy resin were used in 1966. The greater portion of this epoxy was used as an adhesive for concrete, but a significant percentage was used as the binder material in epoxy concrete. Other resins, such as the polyesters, are being actively investigated [18].

Uses of polymer concrete have thus far been small-scale operations, mostly repair work in nonbuilding construction such as bridges. Epoxy resins can be combined with sand and used as a patching material for concrete. Traditionally, concrete patches have been difficult to apply successfully. Usual practice was to cut back the concrete area to a depth of 2 in. Sides of the patch were cut vertically because the patches could not be featheredged. With epoxy concrete, patching is somewhat easier. The surface must be cut back to sound concrete, because any patch is only as good as the substrate to which it is bonded. However, the patch does not have to be any specified depth, and featheredging of patches is feasible with epoxy mortars. When the epoxy mortars were first used as patching materials, frequent failures were noted in the concrete below the bond line. Many of these failures were attributed to the difference in strength and the coefficient of expansion between the portland cement concrete and the epoxy mortar. Subsequent research has developed more flexible epoxy systems which have given excellent results.

Polymer concretes are currently used for the surfacing of bridge decks and audible warning strips ("rumble strips") in high hazard traffic areas, and for providing extra skid resistance in high density intersections. Parking garages with steep ramps can also use these materials to advantage. The most commonly used polymer concrete surfacing is based on epoxy resin. The epoxies used are usually given added flexibility by the addition of either a coal tar or a polysulfide. Usual practice is to use a truck-mounted sprayer to cover one traffic lane width at a time with the resin. Graded aggregate is then spread over the surface to provide skid resistance. After the resin has cured, the excess aggregate may be swept up and reused. This type of application provides a good wearing surface with very little addition of dead load to the structure as compared with other surfacing materials. The George Washington Bridge in New York City now has an epoxy concrete wearing surface.

The polymer concretes offer certain advantages so that their use will spread into other areas of construction. The properties vary, of course, with the type of binder being used. Many different polymers are currently being investigated, including epoxies, polyesters, polysulfides, polyvinyl acetate, and a series of urethanes. Polyester and epoxy concrete give compressive strengths of 12,000 to 20,000 psi, as compared to 3000 to 5000 psi for portland cement concrete. These polymer concretes also offer a fast cure: 72 to 96 hr versus 28 days for normal concrete. The disadvantages of the polymer concretes are the high heat generated during cure and the high cost. Also, a great deal of research must be done on creep and long-term weathering properties before architects and engineers will specify these materials for beams, columns, and slabs. The more resilient binders, such as the polysulfides and urethanes, give low strength concretes which are capable of very large deformations without failure. These materials will probably find application as resilient coatings.

Quite recently, polyester concrete has gone high fashion. Sink and counter tops, vanities, stair treads, tabletops, bathtubs, bar tops, and a host of other items are now fabricated of polyester resin and colored aggregates. The range of colors is virtually unlimited and the material can be custom-molded into almost any shape.

Polymer concretes are also used for beautiful terrazzo floors.

27.2 Composite Beams

The most commonly used type of composite beam consists of a concrete slab resting on the top flange of a steel beam. Some type of shear connector is used to prevent slip between the beam and slab and thus insure that the composite beam acts as a unit. The shear connectors most frequently used are steel studs welded to the top flange of the beam. In recent years, however, considerable research has been directed towards the use of adhesively bonded composite beams. Three general approaches have been used:

1. The slab is precast and then glued to the steel beam with an epoxy adhesive.
2. The top flange of the steel beam is coated with epoxy resin and the concrete slab is cast on top of the uncured epoxy. The system then cures together.
3. The top flange of the beam is coated with an epoxy or polyester adhesive. Graded aggregate, up to $\frac{1}{4}$ in. size, is then embedded in the resin. After the adhesive has cured, the concrete slab is cast and the exposed aggregate provides the shear resistance. This approach is known as the bonded aggregate beam method.

At the present time all of these systems are experimental. However, it is probable that one or more of these systems will become economically feasible in the near future. For example, the bonded aggregate beam could be shop-fabricated and shipped to the job site with a protective liner around the top flange. One of the current problems with composite construction is that once the studs are in place, it is difficult for the steel workers to walk on top of the beams. With the bonded aggregate beams, the steel workers could simply walk on top of the protective liner until the slab was ready to be cast. The liner could then be removed and the slab concrete placed.

27.3 *The Future of Composite Materials*

There are actually very few pure materials used in construction. In the broadest sense of the word most construction materials are composites, which may be either natural or man-made. Wood is really a composite material which consists of a group of roughly parallel cells, held together by an adhesive. Concrete is a man-made composite. Metallurgists are currently investigating high strength composites formed by embedding whisker fibers of one metal in a matrix of another metal. Decorative laminates, such as Formica °, are actually composite materials. Gypsum wallboard, which consists of a gypsum core and paper surfacing, is also a composite material.

° Trademark of Formica Corporation.

The one property that all these composites have in common is that the elements of the composite are held together by some type of adhesive. As the supply of high grade ores and other natural materials diminishes, it is inevitable that more and more composite materials will be used in construction. The use of adhesives as an integral part of these composites can be expected to grow accordingly.

28

Adhesives for Concrete, Cement, and Plaster

Adhesives for concrete, cement, and plaster cover a wide range of construction problems. Cured concrete may be bonded to cured concrete as in the installation of precast traffic buttons to a highway surface. Steel bridge railings may be glued to the concrete surface of a bridge sidewalk. In the case of spalled or deteriorated concrete, the engineer may wish to rebuild the structure to its former line and grade. Spalled or damaged plaster may also be recoated to furnish a new surface. It is often desirable to use a separate cement mortar floor topping over a structural floor. This separate floor topping can be bonded successfully to existing floors of either wood or concrete. These are all different problems and consequently there is no "cure-all" answer. However, the problems can be grouped and the possible solutions listed so that the best possible match can be made.

1. Bonding cured concrete to cured concrete
2. Bonding cured concrete to other materials
3. Bonding new concrete to cured concrete
4. Bonding new concrete or cement mortar to other materials
5. Bonding new plaster to old plaster

In all of these problems something must be glued to something else, and this requires the use of a bonding agent. In the bonding of cured concrete to cured concrete, the adhesive is applied to the mating surfaces. In the case of new concrete over old, the bonding agent may be incorporated into the new concrete, or it may be applied separately to the old concrete before the new concrete is placed. The bonding of cementitious materials can be done with rubber, acrylic, or vinyl emulsions or with epoxies. The epoxies are the highest priced of the group, but have taken over the market for exterior applications and applications where high strength is required. For indoor use, where low to moderate strength is required, any of the other three adhesives may be applied.

The straight epoxy resin adhesives are very strong, but quite brittle. Consequently, the epoxies furnished to the job are given added flexibility by the addition of another, more flexible resin. Polysulfides, for example are often used. The epoxy, as furnished to the job site, is a two-component material which must be mixed before use. The typical epoxy adhesive has a shear strength of 4000 to 5000 psi and a tensile adhesive strength of 2500 to 3000 psi. The epoxy can be applied by brush or spray and has an open time of 15 min to 4 hr, depending on formulation. Cure time is approximately seven days to reach full strength. High pressure is not required in epoxy bonding. All that is required is enough pressure to keep the parts in contact. Figure 28.1 shows the use of an epoxy resin to install a stone-wearing surface on a pedestrian bridge.

The rubber, acrylic, and vinyl emulsions are one–component, moderate strength adhesives. They have a tensile and shear strength of 100 to 200 psi. These adhesives do not weather as well as the epoxies in exposed locations and the vinyls, moreover, have poor water immersion resistance. These materials are available in paste or brush consistency. For indoor applications, these adhesives offer the advantages of no mixing,

Fig. 28.1 (*a*) Epoxy bonding agent used for installing a stone surface over a concrete slab. In this installation the epoxy is used to bond the mortar bed to the structural concrete slab. The epoxy can be seen as the dark area adjacent to the fresh mortar. (*b*) A finished section of the installation shown in (*a*). In this installation the joints were filled with a mortar consisting of fine sand combined with a very flexible epoxy.

(b)

fast clean-up, and economy. They can be applied faster and at half the cost of the epoxies.

28.1 Bonding Cured Concrete to Cured Concrete

There are many applications of this type of bonding. Precast traffic buttons can be glued to a highway. Precast sections of curb can be bonded into place on city streets. Bearing blocks and machine bases can be bonded to industrial floors. Structural repairs to concrete can be made by bonding. (See Fig. 15.2.) The bonding agent used for this application is an epoxy resin. Some polyesters have been used, but these are not as widely accepted as the epoxies. The epoxy resins are available as brush consistency adhesives for surfaces that mate well, or as a gel for nonmating surfaces. The surfaces which are to be bonded must be clean and sound. Loose aggregate, dust, laitance, and curing compounds must be removed from the surfaces. The epoxy resins are furnished in premeasured two-component kits, which must be mixed immediately before using. Tests have shown that the bond between concrete members is almost always higher than the strength of the concrete.

28.2 Bonding Cured Concrete to Other Materials

Handrails for concrete stairways can be bonded into place. Steel handrails can be glued to a bridge sidewalk. Composite beams may be formed by bonding steel beams to precast slabs. Precast slabs can be glued into place on steel purlins to form a concrete roof deck. These applications also use the epoxy resin. The application is similar to bonding

pieces of concrete together. When bonding to steel, the bonding surface
of the steel should be cleaned down to bare metal, preferably by sandblasting or power brushing with a wire brush.

28.3 Bonding New Concrete to Cured Concrete

Patching and rehabilitation of disintegrated concrete has always been a
troublesome problem for the maintenance engineer. If small volumes of
patching are required, an epoxy mortar is probably the best solution.
However, if larger quantities of concrete are involved, a bonding agent
and portland cement concrete will probably prove to be more economi-
cal. If new concrete is to be bonded to cured concrete, the existing concrete must be clean and sound. If the old concrete shows any signs of
disintegration, the surface should be cleaned with chipping hammers
and sandblasted down to sound concrete. All loose aggregate and dust
must be removed from the surface. The epoxy adhesive can then be applied to the surface by brush or spray and the new concrete can be
placed.

In the placement of industrial floors with large floor areas, a separate
concrete topping is sometimes placed on top of the structural floor slab.
These separate toppings were, for years, quite troublesome until the use
of bonding agents became common. For such indoor applications, the
rubber latex, the acrylic, or the vinyl emulsion would be suitable. These
emulsions may be applied directly by brush or spray, or they may be
mixed with a thin mortar and brushed into place. The epoxies may also
be used for this application, but their high strength (and high cost) are
not required. The cost factors are approximately 10 cents per sq ft for
the epoxy and four to five cents for the other bonding agents.

28.4 Bonding New Concrete to Other Materials

New concrete may be bonded to other construction materials such as
steel, stone, wood, and brick. The epoxy is probably the safest choice for
all these bonding jobs, in spite of its higher cost. The epoxy provides
better strength, moisture and salt resistance, and resistance to freeze-
thaw than the other bonding agents provide.

Some increase in the bonding capabilities of new concrete can be obtained by using a bonding agent as an additive to the new concrete. Polyvinyl acetate has been used successfully for this type of bonding [19].
A thin concrete topping (⅝ in.) with a PVA admix was placed over an

existing wood block floor in the Testing Laboratory at the University of Cincinnati. This topping was placed in 1954 and is still giving satisfactory service.

28.5 Bonding New Plaster to Old Plaster

This is an indoor operation which requires a good bond, but demands little with respect to shear strength. The vinyl emulsions based on PVA are suitable for this application. The PVA adhesives have a shear strength which is adequate for plaster bonding. If the plaster is to be exposed to high humidity conditions, the vinyl bonding agents are not suitable. An epoxy adhesive would function better in the high humidity environment.

Bibliography

[1] Oberdick, W. A., *Computerized Joint Movement Criteria*, Proceedings, Midyear Technical Seminar, ASTM Committee C-24, 1968.

[2] Graham, M. D., et al., *New York State Experience with Concrete Pavement Joint Sealers*, Highway Research Record No. 80, Highway Research Board, 1965.

[3] Morosov, N. V., *Joints of Large Panel Buildings and Their Characteristics*, NBRI Report 51C, Norwegian Building Research Institute, Oslo, 1968.

[4] Dalaker, M., *Gaskets in Window Joints*, NBRI Report 51C, Norwegian Building Research Institute, Oslo, 1968.

[5] Tons, E., *A Theoretical Approach to Design of a Road Joint Seal*, Highway Research Board Bulletin 299, Highway Research Board, 1959.

[6] Cook, J. P., *A Study of Polysulfide Sealants for Joints in Bridges*, Highway Research Record No. 80, Highway Research Board, 1965.

[7] Brown, H. R., "Polychloroprene Gaskets," in *Sealants*, Damusis, A., Ed., Reinhold Publishing Co., 1967.

[8] Koppes, W., *Functional Requirements, Standards and Tests for Joint Seals*, NBRI Report 51C, Norwegian Building Research Institute, Oslo, 1968.

[9] Garden, G. K., *Sensible Use of Sealants*, NBRI Report 51C, Norwegian Building Research Institute, Oslo, 1968.

[10] Higgins, J. J., "Butyl and Related Solvent Release Sealants," in *Sealants*, Damusis, A., Ed., Reinhold Publishing Co., 1967.

[11] Bieneman, R. A., "Drying Oil Caulks," in *Sealants*, Damusis, A., Ed., Reinhold Publishing Co., 1967.

[12] Giordano, J. J., *Recommended Practices for Building Joint Analysis with Regard to Sealant Needs*, Proceedings, Midyear Technical Seminar, ASTM Committee C-24, 1969.

[13] Yamasaki, R. S., "Coatings to Protect Concrete against Damage by De-Icer Chemicals, A Literature Review," *Journal of Paint Technology*, Vol. 9, No. 509, June 1967.

[14] Cook, J. P., and R. M. Lewis, *Evaluation of Pavement Joint and Crack*

Sealing Materials and Practices, NCHRP Report No. 38, Highway Research Board, 1967.

[15] Stead, K. A., *Control of Pavement Movements Adjacent to Structures*, Report No. 1, Engineering Experiment Station, University of Mississippi, 1965.

[16] Rohrer, R. B., "Adhesives for Floor Surfacing Materials," in *Adhesives and Sealants in Buildings*, Publication No. 577, Building Research Institute, Washington, D. C., 1959.

[17] Boot, R. J., "Silicone Sealants," in *Sealants*, Damusis, A., Ed., Reinhold Publishing Co., 1967.

[18] Knab, L. I., *Polyester Concrete, Load Rate Variance*, Highway Research Record No. 287, Highway Research Board, 1969.

[19] Howe, R. T., "Polyvinyl Acetate and Portland Cement Mortars," Paper No. 2358, *Journal of the Construction Division*, Proceedings of the American Society of Civil Engineers, February 1960.

Glossary

Abrasion resistance. Resistance to wear resulting from mechanical action on a surface.

Accelerated aging. A set of laboratory conditions designed to produce in a short time the results of normal aging. Usual factors included are temperature, light, oxygen, and water.

Accelerated weathering. A set of laboratory conditions to simulate in a short time the effects of natural weathering.

Accelerator. An ingredient used in small amounts to speed up the action of a curing agent. Sometimes used as a synonym for curing agent.

Acetone. Dimethyl Ketone. A very volatile solvent. Particularly useful for cleaning metal substrates.

Adherend. A body which is held to another body by an adhesive.

Adhesion. The clinging or sticking together of two surfaces. The state in which two surfaces are held together by forces at the interface.

Adhesion, mechanical. Adhesion due to the physical interlocking of the adhesive with the surface irregularities of the substrate.

Adhesion, specific. Adhesion due to molecular forces at the surface.

Adhesive. A substance capable of holding materials together by surface attachment.

Adhesive failure. Type of failure characterized by pulling the adhesive or sealant loose from the adherend.

Adsorption. The action of a body in condensing and holding gases and other materials at its surface.

Aging. The progressive change in the chemical and physical properties of a sealant or adhesive.

Alligatoring. Cracking of a surface into segments so that it resembles the hide of an alligator.

Ambient temperature. Temperature of the air surrounding the object under construction.

Aroclor. Plasticizer used in some sealants. A chlorinated diphenyl.

255

Asbestos. A mineral with a structure of long, fine fibers.

Asphalt. Naturally occurring mineral pitch or bitumen.

Back-up. A compressible material used at the base of a joint opening to provide the proper shape factor in a sealant.

Blown oils. Oils which have had air blown through them to increase viscosity or alter other properties.

Bond (noun). The attachment at an interface between substrate and adhesive, or sealant.

Bond (verb). To join materials together using an adhesive.

Bond breaker. Thin layer of material used to prevent the sealant from bonding to the bottom of the joint.

Bond face. The part or surface of a building component which serves as a substrate for an adhesive or sealant.

Bond strength. The force per unit area necessary to rupture a bond.

Butt joint. A joint in which the structural units are joined to place the adhesive or sealant into tension or compression.

Butyl rubber. A copolymer of isobutene and isoprene. As a sealant it has low recovery and slow cure, but good tensile strength and elongation.

Carbon black. Finely divided carbon formed by the incomplete combustion of natural gas.

Catalyst. Substance added in small quantities to promote a reaction, while remaining unchanged itself.

Caulk (noun). A sealant with a relatively low (less than 20%) movement capability.

Caulk (verb). To fill the joints in a building with a sealant.

Cellular material. A material containing many small cells dispersed throughout the material. The cells may be either open or closed.

Chalking. Formation of a powdery surface due to weathering.

Checking. The formation of slight breaks or cracks in the surface of a sealant.

Chemical cure. Curing by chemical reaction. Usually involves the cross-linking of a polymer.

Closed cell. A cell enclosed by its walls and therefore not connected to other cells.

Coefficient of expansion. The coefficient of linear expansion is the ratio of the change in length per degree to the length at $0°C$.

Cohesion. The molecular attraction which holds the body of a sealant or adhesive together. The internal strength of an adhesive or sealant.

Cohesive failure. The failure characterized by pulling the body of the sealant or adhesive apart.

Compression seal. A preformed seal which is installed by being compressed and inserted into the joint.

Compression set. The amount of permanent set that remains in a specimen after removal of a compressive load.

Cone penetrometer. An instrument for measuring the relative hardness of soft, deformable materials.

Crazing. A series of fine cracks which may extend through the body of a layer of sealant or adhesive.

Creep. The deformation of a body with time under constant load.

Cross-linked. Molecules that are joined side by side as well as end to end.

Cure. To set up or harden by means of a chemical reaction.

Cure time. Time required to effect a complete cure at a given temperature.

Curing agent. A chemical which is added to effect a cure in a polymer.

Depolymerization. Separation of a complex molecule into simpler molecules.

Dowel. A straight steel bar used to transfer load between sections of a concrete pavement slab.

Elasticity. The ability of a material to return to its original shape after removal of a load.

Elastomer. A rubbery material which returns to approximately its original dimensions in a short time after a relatively large amount of deformation.

Emulsion. A dispersion of fine particles in water.

Exothermic. A chemical reaction which gives off heat.

Extender. An organic material used to increase the volume and lower the cost of a sealant or adhesive.

Extensibility. The ability of a sealant to stretch under tensile load.

Extrusion failure. Failure which occurs when a sealant is forced too far out of the joint. The sealant may be abraded by dirt or folded over by traffic.

Fatigue failure. Failure of a material due to rapid cyclic deformation.

Field Molded sealant. A mastic sealant which takes its shape by being placed into the joint.

Filler. Finely ground material added to a sealant or adhesive to change or improve certain properties.

Flashing. Strips, usually of sheet metal, to waterproof the junctions of building surfaces, such as roof peaks and valleys, and the junction of a roof and chimney.

Gasket. A deformable material placed between two surfaces to seal the union between the surfaces.

Gypsum wallboard. A sandwich type material. Gypsum plaster with a heavy paper coating on both sides. When fastened directly to studs, it forms a wall surface.

Hardboard. Fine pieces of wood bound together with an adhesive and pressed into 4- $\times$ 8-ft sheets. Thickness is approximately $\frac{1}{8}$ in. Thermosetting resins are usually used as the adhesive binder.

Hardener. A substance added to control the reaction of a curing agent in a sealant or adhesive. Sometimes used as a synonym for curing agent.

Hardness. The resistance of a material to indentation. On the Shore A scale, hardness is measured in relative numbers from 0 to 100.

Head. The top member of a window or door frame.

Interface. The common boundary surface between two substances.

Jamb. The side of a window, door opening, or frame.

Joint (adhesive use). The point at which two substrates are joined by an adhesive.

Joint (sealant use). The opening between component parts of a structure.

Laitance. A thin, weak coating which sometimes forms on the surface of concrete.

Lap joint. A joint in which the component parts overlap so that the sealant or adhesive is placed into shear action.

Latex (natural). Milky juice of the rubber plant. A natural rubber emulsion.

Latex caulks. A rubbery emulsion caulking material. The most common latex caulks are polyvinyl acetate or vinyl acrylic.

Load transfer device. Any device embedded in the concrete on both sides of a pavement joint to prevent relative vertical movement of slab edges.

Mastic (broad interpretation). Any field molded sealant or adhesive. Includes materials which are gunned, poured, or troweled into place.

Mastic (narrow interpretation). A thick, pasty sealant or adhesive.

Mercaptan. An organic compound containing —SH groups.

Mil. One—thousandth of an inch.

Modulus. The ratio of stress to strain.

Monomer. A material composed of single molecules. A building block in the manufacture of polymers.

Mullion. External structural member in a curtain wall building. Uusually vertical. May be placed between two opaque panels, between two window frames, or between a panel and a window frame.

Neck down. The change in the cross sectional area of a sealant as it is extended.

Oil (drying). Oils which dry to a hard varnish-like film. Linseed oil is a common example.

Open cell. A cell which is interconnected to other cells or to the surface of the body.

Open time. Time interval between when an adhesive is applied and when it becomes no longer workable.

Oxidation. Formation of an oxide. Also the deterioration of rubbery materials due to the action of oxygen or ozone.

Ozone. A reactive form of oxygen. A powerful oxidizing agent, it occurs naturally in the atmosphere.

Particle board. Same as hardboard except that larger wood chips are used as the filler.

Pavement growth. Increase in the length of a pavement caused by incompressible material in the joints.

Peel test. A test of an adhesive or sealant using one rigid and one flexible substrate. The flexible material is folded back (usually 180°) and the substrates are peeled apart. Strength is measured in pounds per inch of width.

Permanent set. The amount of deformation which remains in a sealant or adhesive after removal of a load.

Phenolic resin. A thermosetting resin. Usually formed by the reaction of a phenol with formaldehyde.

Pigment. A coloring agent added to a sealant.

Pitch. The residue which remains after the distillation of oil, and so forth, from raw petroleum.

Plasticizer. A material which softens a sealant or adhesive by solvent action.

Poise. The cgs unit of viscosity. Example: A polysulfide highway joint sealant might have a viscosity of 500 poises, at 77 F. Higher numbers indicate a more viscous material.

Polymer. A compound consisting of long chain-like molecules. The building units in the chain are monomers.

Polysulfide rubber. Synthetic polymer usually obtained from sodium polysulfide. The polymer segments are generally terminated with —SH groups. Polysulfide rubbers make very good sealants.

Pot life. (*See* Working life.)

Preformed sealant. A sealant which is preshaped by the manufacturer before being shipped to the job site.

Pressure sensitive adhesive. Adhesive which retains tack after release of the solvent, so that it can be bonded by simple hand pressure.

Primer. A preparatory material which is applied to joint faces in order to improve adhesion.

Reflection cracks. A crack through a bituminous overlay on a portland cement concrete pavement. The crack occurs above any working joint in the base pavement.

Reinforcement (in rubbers). Increase of modulus, toughness, tensile strength, and so forth, by the addition of selected fillers.

Resilience. A measure of energy stored and recovered during a loading cycle. It is expressed in percent.

Resins. Solid organic materials, generally not soluble in water, which have little or no tendency to crystallize. Example: Epoxy and polyester resins.

Retarder. A substance (generally a mild acid) added to slow down the cure rate of a sealant or adhesive.

Routing. Removing old sealant from a joint by means of a rotating bit.

Rubber latex. Water emulsion of an elastomer.

Sealant. Any material used to seal joints or openings against the intrusion or passage of any foreign substance such as water, gases, air, or dirt.

Sealer. A surface coating generally applied to fill cracks, pores, or voids in the surface.

Self-leveling sealant. A sealant which is fluid enough to be poured into horizontal joints. It forms a smooth, level surface without tooling.

Shape factor. The width to depth proportions of a field molded sealant.

Shear test. A method of deforming a sealed or bonded joint by forcing the substrates to slide over each other. Shear strength is reported in units of force per unit area (psi).

Shelf life. The length of time a sealant or adhesive can be stored and still retain its properties.

Shrinkage. Percentage weight loss under specified conditions.

Silicone rubber. A synthetic rubber based on silicon, carbon, oxygen, and hydrogen. Silicone rubbers are widely used as sealants and coatings.

Skewed joints. Transverse joints in a pavement slab, which are placed at an angle and not perpendicular to the direction of traffic.

Solvent. Liquid in which another substance can be dissolved.

Spalling. A surface failure of concrete, usually occurring at the joint. It may be caused by incompressibles in the joint, by overworking the concrete, or by sawing joints too soon.

Sponge. Cellular material, usually of open cell construction.

Strain. Deformation per unit length. Example: Change in length divided by the original length of a test specimen. Strain is a dimensionless number. Small strains are expressed in units of inches per inch. Strains in rubbery materials may be expressed in percent.

Stress. Force per unit area, usually expressed in pounds per square inch (psi).

Stress relaxation. Reduction in stress in a material which is held at a constant deformation for an extended time.

Subgrade. The earth or granular fill below a pavement slab.

Substrate. An adherend. The surface to which a sealant or adhesive is bonded.

Tackiness. The stickiness of the surface of a sealant or adhesive.

Tear strength. The load required to tear apart a sealant specimen. ASTM test method D-624 expresses tear strength in pounds.

Tensile strength. Resistance of a material to a tensile force (a stretch). The cohesive strength of a material, expressed in psi.

Thermoplastic. A material which can be repeatedly softened by heating. Thermoplastics generally have little or no chemical cross-linking.

Thermosetting. A material which hardens by chemical reaction. Not remeltable. The reaction usually gives off heat.

Thixotropic. Nonsagging. A material which maintains its shape unless agitated. A thixotropic sealant can be placed in a joint in a vertical wall and will maintain its shape without sagging during the curing process.

Toxic. Poisonous or dangerous to humans by swallowing, inhalation, or contact resulting in eye or skin irritation.

Transverse joint. A joint perpendicular to the direction of traffic in a highway pavement.

Ultimate elongation. Elongation at failure.

Ultraviolet light. Part of the light spectrum. Ultraviolet rays can cause chemical changes in rubbery materials.

Urethane. A family of polymers ranging from rubbery to brittle. Usually formed by the reaction of a diisocyanate with a hydroxyl.

Vehicle. The liquid component of a material. Example: Oil paint is composed of a vehicle (linseed oil) and pigment.

Viscosity. A measure of the flow properties of a liquid or paste. Example: Honey is more viscous than water. Water (the standard of comparison) has a viscosity of $1/100$ of a poise.

Vulcanization. Improving the elastic properties of a rubber by a chemical change

Weatherometer. An environmental chamber in which specimens are subjected to water spray and ultraviolet light.

Working life. Period of time after mixing, during which a sealant or adhesive can be used.

Sources Used in Preparation of of this Glossary

American Concrete Institute, Committee Report 504.

Gay, C. M., and H. Parker *Materials and Methods of Architectural Construction,* John Wiley & Sons, 1943.

Highway Research Board, Committee MC-D3, *Glossary of Sealant Terminology.*

Huntington, W. C., *Building Construction,* John Wiley & Sons, 1964.

Index